Serotonin

Edited by Ying Qu

Published in London, United Kingdom

IntechOpen

Supporting open minds since 2005

Serotonin
http://dx.doi.org/10.5772/intechopen.72010
Edited by Ying Qu

Contributors
Adela Hilda Onutu, Cristina Petrisor, Dan Sebastian Dîrzu, Katy Satué Ambrojo, Sara Puig-Perez, Ying Qu

First published in London, United Kingdom, 2019 by IntechOpen
IntechOpen is the global imprint of INTECHOPEN LIMITED, registered in England and Wales, registration number: 11086078, The Shard, 25th floor, 32 London Bridge Street
London, SE19SG - United Kingdom
Printed in Croatia

British Library Cataloguing-in-Publication Data
A catalogue record for this book is available from the British Library

Additional hard copies can be obtained from orders@intechopen.com

Serotonin
Edited by Ying Qu
p. cm.
Print ISBN 978-1-78985-235-6
Online ISBN 978-1-78985-236-3

For EU product safety concerns:
IN TECH d.o.o., Prolaz Marije Krucifikse Kozulić 3, 51000 Rijeka, Croatia, info@intechopen.com or visit our website at intechopen.com.

Meet the editor

Dr. Ying Qu is a multi-disciplinary scientist, currently working in Leulan Bioscience, USA. She received her BS and MS in Chemistry from Lanzhou University, China and her PhD in Neuroscience from the Catholic University of Leuven, Belgium. Dr. Qu has spent part of her career at the National Institutes of Health, USA, studying depression mechanisms underlying serotonin post-receptor regulated signaling transduction. She is also involved in a drug discovery program at Johnson and Johnson in the USA developing novel dual-acting antidepressants with selective serotonin reuptake inhibitors. In 2002, she received a Sevier Young Investigator Award from the Serotonin Club at the International Union of Basic and Clinical Pharmacology (IUPHAR) Satellite Meeting on Serotonin. She has published over 30 peer-reviewed papers, 40 abstracts and two book chapters in the fields of neuropsychopharmacology and bioanalysis.

Contents

Preface

Serotonin is a monoamine neurotransmitter in the central nervous system (CNS), whose well-known biological functions include modulating cognition, sleep, emotion, learning, memory, and numerous physiological processes.

At any given time, over 4% of the global population suffers from a major depressive disorder. Among approved depression treatments are selective serotonin reuptake inhibitors (SSRIs). These are based on the serotonin hypothesis, which holds that low levels of extracellular serotonin causes depression; consequently, increasing extracellular serotonin can treat depression. Since the introduction of SSRIs, many books about serotonin have been published.

I started my neuroscience career by measuring serotonin in the brain of different animals during my PhD study. Later on, I became involved in depression mechanisms research and drug discovery. This scientific journey brought me from Lanzhou in China, Leuven in Belgium, Bethesda in the USA, all the way to San Diego.

Sometimes I pick up an assortment of scattered seashells while walking along the beautiful Torrey Pines Beach in San Diego. Likewise, this book contains an assortment of discussions of different aspects of serotonin to enrich our knowledge and understanding of this neurochemical.

Ying Qu, PhD
Leulan Bioscience,
San Diego, CA

Chapter 1

Introductory Chapter: From Measuring Serotonin Neurotransmission to Evaluating Serotonin Post-Receptor Signaling Transduction

Ying Qu

1. Introduction

Serotonin or 5-hydroxytryptamine (5-HT) is a well-established monoamine neurotransmitter in the central nervous system (CNS). The discovery of 5-HT dates as far back as 1868 and can be traced to its presence in the blood and in the gastrointestinal tract [1]. Its well-known biological functions include modulating cognition, sleep, emotion, learning, memory, and numerous physiological processes. 5-HT is primarily found in the enteric nervous system located in the gastrointestinal tract [2], where it

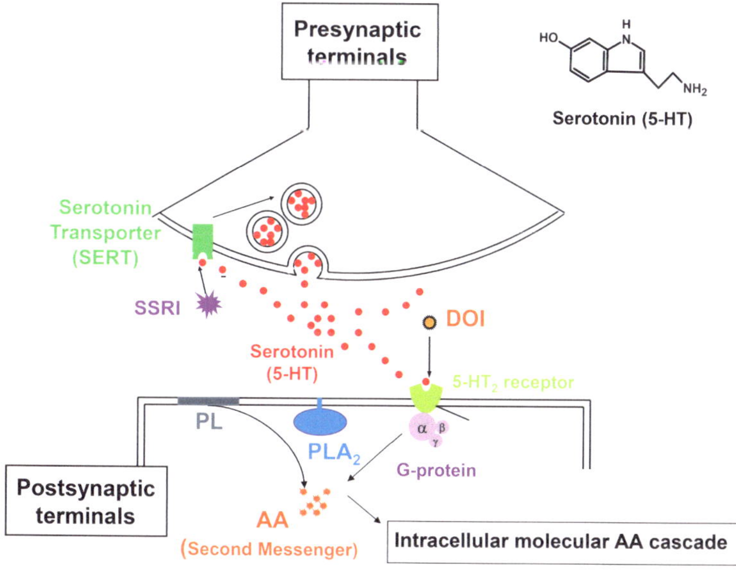

Figure 1.
Model explaining PLA_2 activation in response to serotonergic drugs. Under normal conditions, the 5-HT that is released from presynaptic vesicles into the synaptic cleft binds to postsynaptic 5-HT receptors coupled via a G-protein to PLA_2, thus hydrolyzing arachidonic acid (AA) from membrane phospholipids (PL). Administration serotonergic drugs activate PLA and increase incorporation of AA by different routes. (1) 5-$HT_{2A/2C}$ agonist, DOI directly binds to 5-HT_2 receptors to activate this signal; (2) fluoxetine (SSRI) inhibits 5-HT uptake, thus increasing 5-HT in the synaptic cleft so as to increase PLA activation and AA release. This figure adapted from [23].

regulates intestinal movements [2], and the remainder is synthesized in the serotonergic neurons of the CNS, where it has various functions such as the regulation of mood, appetite, and sleep. Modulation of 5-HT at synapses is thought to be a major action of several classes of pharmacological antidepressants. Among these, selective serotonin reuptake inhibitors (SSRIs), such as fluoxetine and citalopram, are the most important class of antidepressant in the treatment of major depressive disorder (MDD) and anxiety disorders [3]. The exact mechanism of action of SSRIs is not fully revealed. SSRIs are able to increase the extracellular level of the neurotransmitter 5-HT by inhibiting its reuptake into the presynaptic terminal, increasing the level of 5-HT in the synaptic cleft available to bind to the postsynaptic 5-HT receptor (as shown in **Figure 1**). SSRIs have different degrees of selectivity for the other monoamine transporters, and the most selective SSRI has weak affinity for the norepinephrine and dopamine transporters. They are the most widely prescribed antidepressants in many countries, and their efficacy in mild or moderate cases of depression has been disputed [4] and may be outweighed by side effects [3]. I have been involved in 5-HT research for two decades. This chapter summarized my research on 5-HT-related projects from measuring 5-HT concentration, attempting to discover a new generation of SSRIs to investigate 5-HT-regulated post-receptor signaling transduction. This chapter also discusses some perspectives research that is important for SSRI and depression treatment.

2. Measuring serotonin in CNS system

In the early 1990s, liquid chromatography (LC) with an electrochemical detector (ED) had been widely used for the measurement of neurochemicals [5]. The first 5-HT project that I worked with was to develop a method for measuring 5-HT concentration in chicken brain tissue [6]. An isocratic LC-ED for the determination of L-3,4-dihydroxyphenylalanine, dopamine, norepinephrine, epinephrine, 5-HT, and their major metabolites, 3,4-dihydroxyphenylacetic acid, 4-hydroxy-3-methoxyphenylacetic acid, and 5-hydroxyindole-3-acetic acid in chicken brain tissue was developed in our lab. The method was applied to study the influence of food restriction on the concentration of 5-HT and other monoamine neurotransmitters in different brain areas, known to be involved in the feeding and reproductive behavior of female broiler chickens. In the experiment, two to six micropunches from 20 different brain areas on 300 μm cryostat brain section were punched out and expelled into Eppendorf for homogenization and extraction. Supernatant was injected onto LC-ED, and over 1000 micro-punched tissue samples from ad libitum fed and food-restricted female broiler chickens were analyzed. Tissue pellets were dissolved in PBS buffer for protein content determination to express the results as pg monoamine/μg protein. Although the concentration of monoamines in the brain is not high, multiple tissue micropunches made enough amount of monoamine and 5-HT to match the sensitivity of the assay. Our results provided a possible role for catecholamines and indolamines in the altered feeding and reproductive behavior of the broiler chicken [6]. To finish my Ph.D. thesis, I modified this method to measure 5-HT and other monoamine neurotransmitters in cat visual cortex [7]. The role of monoaminergic neuromodulators in the reorganization of cortical topography following limited sensory deprivation in the adult cat was investigated in this study [8]. The total concentrations of dopamine, noradrenaline, 5-HT, and their major metabolites were measured in the visual cortex of both control and experimental animals using this microbore LC-ED method. The sensory deprivation cats were subjected to a binocular retinal lesion corresponding to the central 10 degrees of vision and sacrificed 2 weeks post-lesion. The deprivation was confirmed in area 17 by measuring immediate-early gene if-268 messenger RNA expression. The total concentration of 5-HT was significantly lower in the deprived cortex, and the metabolite of 5-HT, 5-hydroxyindole-3-acetic acid, was significantly higher in the nondeprived cortex than in deprived cortex and normal cortex. The levels of noradrenaline and dopamine were significantly higher in the nondeprived cortex of retinal lesion cats than in the deprived cortex of retinal lesion cats and the cortex of normal animals. This pattern follows the release of the excitatory neurotransmitter glutamate under the same conditions. These results suggest that the modulation of 5-HT, noradrenaline, and dopamine is regulated by visual afferent activity [8].

To switch my scientific career to the pharmaceutical industry, I joined the CNS drug discovery team for making a new generation dual function SSRI [9] for depression treatment. Fluoxetine (Prozac) [10] is the first SSRI and widely used for the treatment of depression which was used as reference compounds for new SSRI discovery. Fluoxetine exerts its behavioral and clinical therapeutic effect by blocking the transport of 5-HT at the serotonin reuptake transporter (SERT), thereby increasing extracellular level of 5-HT in the serotonergic synaptic cleft of many brain regions as shown in **Figure 1**. In vivo microdialysis has been extensively used to document the changes of extracellular level of 5-HT in the rat brain after administration of fluoxetine [11]. Therefore, we designed a 21-hour in vivo microdialysis experiment and the effect of acute systemic administration of fluoxetine (3 and 10 mg/kg s.c.) on extracellular level of 5-HT in the frontal cortex of freely moving rats was analyzed by LC with ESA CoulArray coulometric detector (an electrochemical detector) [9, 12]. In this experiment, the guide cannula was implanted on rats' brain by surgery and secured in place with skull screws and dental cement. Animals were allowed at least 3 days to recover from surgery prior to experimentation. Dialysis probes were perfused with artificial cerebral spinal fluid (aCSF, 47 mM NaCl, 4 mM KCl, 0.85 mM $MgCl_2$, 2.3 mM $CaCl_2$, pH 7.4) at a flow rate of 1 μL/min. Samples were collected every 60 min. Microdialysates were analyzed by LC-ED. Separation was performed on a C18 column. All values for microdialysis studies were calculated as percentage change at each time point compared with the average of three baseline values. Due to the limitation of low recovery of microdialysis probe (less than 20% in average) and low concentration of 5-HT in the frontal cortex of rat brain (about 100 fg/μL in this microdialysates), high sensitivity analytical tool is required. LC-ED was the most popular method to measure 5-HT. In recent years, liquid chromatography with tandem mass spectrometry (LC-MS/MS) was also used for this purpose [13].

Pharmacokinetic (PK) characterization and in vivo pharmacological properties of new chemical entities are important components during lead compound selection and optimization in the drug discovery process. Accordingly, reliable techniques are needed that can generate the requisite pharmacokinetic/pharmacodynamic (PK/PD) information for an increased number of compounds. When dealing with compounds targeting the central nervous system (CNS), biophase PK may differ significantly from plasma PK, because blood-brain barrier (BBB) transport and brain distribution often do not occur instantaneously and to a full extent. In vivo microdialysis technique can be used to collect not only the extracellular endogenous substances but also the extracellular free drug in the same local interstitial environment, which may reflect the amount of drug available at the pharmacological target. However, the application of this technique was highly limited by the lack of the proper sensitive analytical methods to determine the endogenous substance and exogenous drug. LC-MS/MS technique improvement provides a direct, structural-specific measurement of individual components with very high sensitivity. The mass spectrometer has minimal baseline drift and can be equilibrated very rapidly. For this purpose, we have developed a series of LC-MS/MS methods, which enable us to monitor drug, citalopram, and 5-HT in the same microdialysis samples [13]. These applications demonstrated in vivo microdialysis coupled with LC-MS/MS is a very important tool to evaluate the PK/PD relationship by comparing the time course of free drug versus biomarker. LC-MS/MS method measuring 5-HT concentration in the brain is possible, but not widely applied [13].

3. Evaluating PK/PD profile of the dual function SSRI

The World Health Organization (WHO) estimates that more than 300 million individuals of all ages suffer from depression [14]. SSRIs have been the drugs for depression treatment. These drugs increase 5-HT levels in the synaptic cleft by inhibiting its reuptake into the presynaptic neuron through blockade of the SERT. Although many patients experience relief after treatment with one of the many marketed SSRIs, efficacy is noticeable only after weeks of treatment. Many physicians are reported to co-prescribe stimulants with SSRI to provide subjective relief during the beginning weeks of antidepressant therapy [15]. Most of these stimulants are increased dopamine release and produced robust behavioral activation, which had the risk of allowing patients to act on their suicidal ideation. It is very important to choose other classes of molecules that have been shown to produce wakefulness in animals without releasing dopamine or producing behavioral activation. Wake-promoting agents such as modafinil are used in the clinic as adjuncts to antidepressant therapy in order to alleviate lethargy. Histamine H_3 receptor antagonist has been demonstrated having the wake-promoting action in numerous animal studies and may therefore be a viable strategy for use as an antidepressant therapy in conjunction with SSRIs. Therefore, some potential antidepressant molecules were created, which combined the wake-promoting effect of a histamine H_3 receptor antagonist with 5-HT reuptake blockage effects of SERT inhibitor [9]. The synthetic approach and structure-activity relationships associated with this effort have been studied [16–18]. In vivo microdialysis experiments were used to examine whether a compound was capable of inducing a robust and persistent increase in 5-HT level over baseline. One of these molecules, JNJ-28583867 (2-methyl-4-(4-methylsulfanyl-phenyl)-7-(3-morpholin-4-yl-propoxy)-1,2,3,4-tetrahydro-isoquinoline), is a selective and potent histamine H_3 receptor antagonist (Ki = 10.6 nM) and inhibitor of the SERT (Ki = 3.7 nM), with 30-fold selectivity for SERT over the dopamine and norepinephrine transporters [9]. After subcutaneous administration, JNJ-28583867 significantly increased cortical extracellular levels of 5-HT as shown in **Figure 2A**. Baseline measurements of 5-HT levels were performed for 4 h prior to administration of JNJ-28583867. At all doses, 5-HT levels remained elevated for the duration of the experiment up to 18 h after dosing. JNJ-28583867 was also tested in a classical test of antidepressant activity, the mouse tail suspension model. As was expected based on the neurochemical profile of JNJ-28583867, an increase in struggling time was observed. Some PK characterization of JNJ-28583867 was carried out in the rat. The behavioral experiments had indicated good oral bioavailability and this was confirmed. The half-life correlates well with the observation that effects could be observed up to 24 h after a single oral dose, as was the case in the head twitch test. The plasma and brain levels of JNJ-28583867 are sustained and correlated reasonably well with efficacy for an extended period of time as shown in **Figure 2B** [9]. Similar PK/PD profiles were observed from norfluoxetine, which is the metabolite of reference SSRI, fluoxetine [12]. Norfluoxetine is the most important active metabolite of the widely used antidepressant fluoxetine. Following subcutaneous administration of fluoxetine in rats, plasma, and brain PK of fluoxetine and norfluoxetine were monitored, respectively, by LC-MS/MS. The extracellular level of 5-HT in the frontal cortex was measured by microdialysis as a PD endpoint. Norfluoxetine when directly

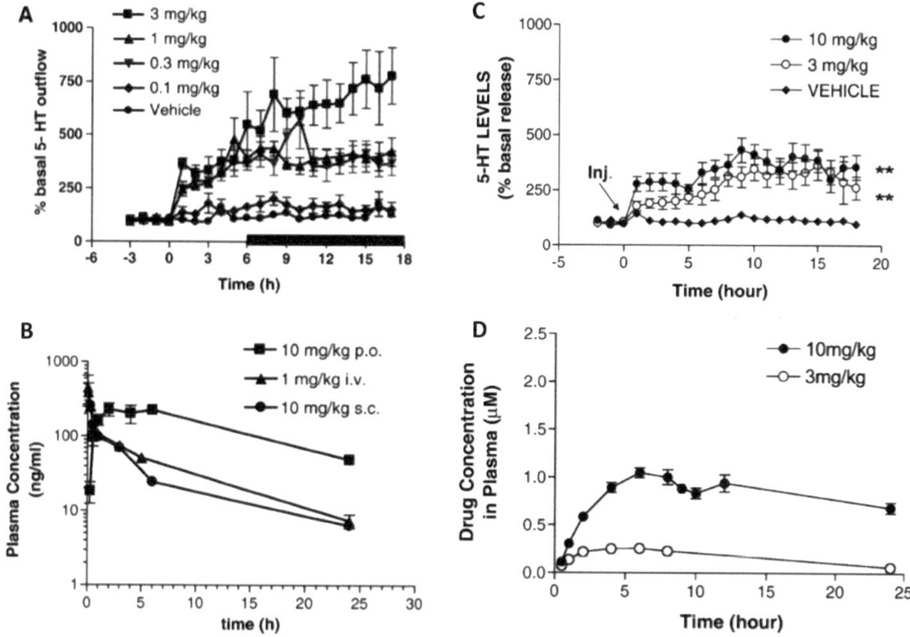

Figure 2.
(A). Effect of JNJ-28583867, administered s.c., on extracellular 5-HT levels in the frontal cortex of male Sprague-Dawley rats. Microdialysis time course. Results are expressed as the average ± S.E.M. of n = 3–6 rats per group. (B). Plasma levels of JNJ-28583867 after oral (10 mg/kg, square), intravenous (1 mg/kg, triangle), and subcutaneous (10 mg/kg, circle) administration to the rat. Results are shown as the average ± S.D. of n = 2–3 samples. (C). Effect of norfluoxetine on the extracellular level of 5-HT in the frontal cortex of free moving rat. Values are mean ± S.E.M. of extracellular 5-HT levels and expressed as a percentage of the average of three baseline samples (defined as 100%). Two-way ANOVA-post-hoc Duncan's multiple range tests were used for comparison. Control (n = 5), 3 mg/kg (n = 6), and 10 mg/kg (n = 6) norfluoxetine were subcutaneously administrated. Asterisks indicate significance of overall effect of drug treatment versus vehicle, P < 0.01. (D). Time course of plasma concentrations of fluoxetine and norfluoxetine. Plasma concentrations (mean ± S.E., n = 10) of norfluoxetine were measured following subcutaneous administration of 3 or 10 mg/kg fluoxetine. ***Figure 4A** and **B** was adapted from [9];* ***Figure 4C** and **D** was adapted from [12].*

administrated to rats caused a significant increase in the extracellular level of 5-HT in the frontal cortex and maintained for 18 hours as shown in **Figure 2C**. This result is correlated well with higher plasma and brain concentration and longer plasma and brain retention time of norfluoxetine (as shown in **Figure 2D**) [12]. In summary, these studies have shown that the combination of histamine H_3 receptor antagonism with SSRI activity in a single molecule results in a pharmacology consistent with the combination of either class of molecule alone. JNJ-28583867 can be a prototype of such a compound to improve current SSRI efficacy and safety profiles [9].

4. Serotonin-mediated post-receptor signaling transduction

Although antidepressants are generally effective in the treatment of MDD, side effects still exist. Serotonin syndrome is a potentially life-threatening adverse drug reaction that results from therapeutic drug use and a predictable consequence of excess serotonergic agonism of CNS and peripheral serotonergic receptors [19]. In 2002, the Toxic Exposure Surveillance System, which receives case descriptions from office-based practices, inpatients settings, and emergency department, reported 26,733 incidences of exposure to SSRIs that caused significant toxic effects in 7349 persons and resulted in 93 deaths [19, 20]. The development mechanism of serotonin syndrome is unknown. It is hypothesized that the level of 5-HT elevation in blood plasma has to be 10–15% above the baseline levels to result in 5-HT toxicity [21]. Several lines of evidence converge to suggest that agonism of 5-HT_{2A} receptors contributes substantially to the condition [22].

To address this question, we studied 5-HT-mediated post-receptor signaling transduction [23]. The 5-HT_2 receptor is G protein-coupled receptor and is recognized to be coupled to the phospholipase A_2 (PLA_2) signaling pathway, stimulating the release of the second messenger, arachidonic acid (AA). This signaling pathway is illustrated in **Figure 1**. PLA_2 activation can be initiated by serotonergic 5-HT_2 receptors via a G-protein. The in vivo fatty acid methods were developed in our lab to measure regional brain incorporation of a radiolabeled fatty acid, including [5,6,8,9,11,12,14,15-^{3}H] arachidonic acid (^{3}H-AA) in conscious rats. Tracer incorporation, represented as the incorporation coefficient k*, reflects PLA_2-mediated AA release. Activation of PLA_2 in the brain is revealed as increments in k* in different receptors or to change serotonergic neurotransmission (**Figure 1**). The fatty acid method can be used to evaluate serotonergic neurotransmission mediated by PLA_2 in awake rats. It can quantify and localize brain PLA_2 signaling in response to different drugs administered acutely or chronically.

In rats, 2,5-dimethoxy-4-iodophenyl-2-aminopropane (DOI), which is a 5-$HT_{2A/2C}$ receptor agonist, provokes head twitches, skin jerks, and forepaw tapping, behaviors that are considered part of a "5-HT syndrome" [24]. The responses usually appear at a dose of 1.0 mg/kg and peak at 2.5 mg/kg. In one of our studies, DOI, when administered to unanesthetized rats, produced widespread and significant increases, of the order of 60%, in k* for arachidonate, particularly in neocortical brain regions reported to have high densities of 5-HT_{2A} receptors [25]. The increases could be entirely blocked by chronic pretreatment with mianserin, a 5-HT_2 receptor antagonist, which is an atypical antidepressant [25]. The results suggest that the 5-HT_2 syndrome involves widespread brain activation of PLA_2 via 5-HT_{2A} receptors, leading to the release of the second messenger, arachidonic acid. Chronic mianserin, a 5-HT_2 antagonist, prevents this activation [25]. In another study, brain PLA_2-mediated signal transduction in response to acute fluoxetine administration in unanesthetized rats had been imaged [26]. By inhibiting presynaptic 5-HT reuptake, fluoxetine is thought to act by increasing 5-HT in the synaptic cleft, thus 5-HT binding to postsynaptic 5-$HT_{2A/2C}$ receptors, activates PLA_2 pathway, and releases the second messenger AA from synaptic membrane phospholipids. To image this activation, fluoxetine (10 mg/kg) or saline vehicle was administered i.p. to unanesthetized rats, and regional brain incorporation coefficients k* of intravenously injected radiolabeled AA were measured after 30 min. Compared with vehicle, fluoxetine significantly increased k* in prefrontal, motor, somatosensory, and olfactory cortex, as well as in the basal ganglia, hippocampus, and thalamus. Many of these regions demonstrate high densities of the SERT and of 5-$HT_{2A/2C}$ receptors. The brain stem, spinal cord, and cerebellum, which showed no significant response to fluoxetine, have low densities of the transporters and receptors. The results show that it is possible to image quantitatively PLA_2-mediated signal transduction in vivo in response to fluoxetine [26]. Fluoxetine's therapeutic action when chronically administered has been ascribed to desensitization of pre-synaptic 5-HT_{1A} and 5-HT_{1B} auto-receptors, further augmenting extracellular 5-HT [27]. We thereby conducted a study to see if this signaling process in rat brain would be altered by chronic administration of fluoxetine followed by 3 days of washout of this SSRI [28]. [^{3}H] AA was intravenously injected in unanesthetized rats and used quantitative autoradiography to determine the incorporation coefficient k* for AA (regional brain radioactivity/integrated plasma radioactivity), a marker of PLA_2 activation, in each of 86 brain regions. k* was measured following acute i.p. saline or DOI (1.0 mg/kg i.p.), in rats injected for 21 days with 10 mg/kg i.p. fluoxetine or saline daily, followed by 3 days without injection. As shown in **Figure 3**, acute DOI produced statistically significant increments in k* in brain regions with high densities of 5-$HT_{2A/2C}$ receptors, but the increments did not differ significantly between the chronic fluoxetine- and saline-treated rats. Additionally, chronic fluoxetine is compared with saline widely and significantly increased baseline values of k*. These results suggest that 5-$HT_{2A/2C}$ receptor-initiated AA signaling is unaffected by chronic fluoxetine plus 3 days of washout in the rat, but that baseline AA signaling is nevertheless upregulated. This upregulation likely occurs because of significant active drug in the brain, considering the long brain half-lives of its metabolite, norfluoxetine [12]. To further understand SERT regulate brain serotonergic transmission and its mediated signaling transduction, we measured PLA_2 activation in SERT knockout mice (SERT−/−) and their littermate controls (SERT+/+). Following administration of 1.5 mg/kg s.c. DOI to unanesthetized mice injected intravenously with radiolabeled AA, PLA_2 activation, represented as the regional incorporation coefficient k* of AA, was determined with quantitative autoradiography in each of 71 brain regions. As shown in **Figure 4**, in SERT+/+ mice, DOI significantly increased k* in 27 regions known to have 5-$HT_{2A/2C}$ receptors, including the frontal, motor, somatosensory, pyriform and cingulate cortex, white matter, nucleus accumbens, caudate putamen, septum, CA1 of the hippocampus, thalamus, and hypothalamus. In contrast, DOI did not increase k* significantly in any brain region of SERT−/− mice. Head twitches following DOI, which also were measured, were robust in SERT+/+ mice but were markedly attenuated in SERT−/− mice. These results show that a lifelong elevation of the synaptic 5-HT concentration in SERT−/− mice leads to downregulation of 5-$HT_{2A/2C}$ receptor-mediated PLA_2 signaling via AA and of head twitches, in response to DOI. Compared with wild-type mice, DOI-induced k* increments were reduced in

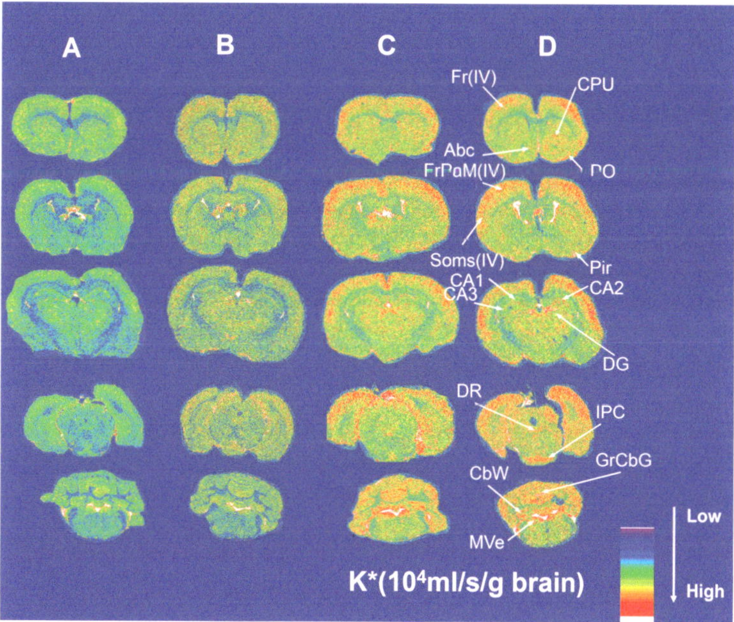

Figure 3.
Coronal autoradiographs demonstrating arachidonic acid incorporation coefficients k. Brain of (A) control rat given acute saline 3 days after receiving i.p. saline for 21 days; (B) control rat given acute DOI (1.0 mg/kg i.p.), 3 days after receiving i.p. saline for 21 days; (C) rat given fluoxetine (10 mg/kg i.p. daily) for 21 days, followed by 3 day washout, and then i.p. Saline on day 24; (D) rat given fluoxetine (10 mg/kg i.p. daily) for 21 days, followed by 3 day washout, and then acute DOI (1.0 mg/kg i.p.). k is color-coded. Abbreviations: Fr (IV), frontal cortex, layer IV; FrPaM (IV), frontal motor (layer IV); Soms, somatosensory cortex; IPC, interpeduncular nucleus; CPU, caudate putamen; CA1, CA2, CA3, DG, regions of the hippocampus; Pir, pyriform cortex; PO, olfactory cortex; GrCbG, granular layer, cerebellar gray; CbW, cerebellar white; DR, dorsal raphe; MVe, medial vestibular nucleus; Abc, nucleus accumbens. This figure adapted from [28].

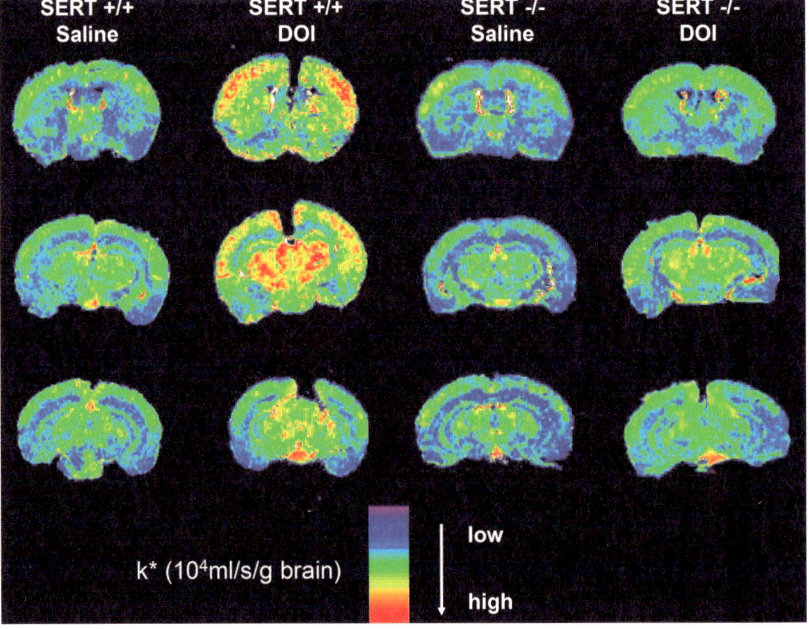

Figure 4.
Coronal autoradiographs demonstrating incorporation coefficients k for arachidonic acid, from brain of SERT+/+ mouse given saline s.c.; SERT +/+ mouse given DOI (1.5 mg/kg s.c.); SERT−/− mouse given saline; SERT+/+ mouse given DOI. k* is color coded. This figure adapted from [29].*

SERT knock out mice [29], but there was no significant effect of 3 weeks of fluoxetine plus washout on DOI-induced k* increments in compared with baseline of chronic fluoxetine treated rats. The difference suggests that a life-long, but not a 3-week, elevation of synaptic 5-HT will downregulate 5-$HT_{2A/2C}$ receptor signaling involving PLA_2.

In summary, these studies suggest that labeled AA can be used to examine in vivo brain PLA_2 signaling initiated by a serotonergic drug. Eventually, brain 5-$HT_{2A/2C}$-mediated signaling coupled to PLA_2 might be imaged in such subjects with positron emission tomography [30].

5. Monitoring therapeutic SSRI in patients

Depression is among the most prevalent psychiatric disorders with a highly variable treatment response and up to one-third of patients not achieving response [31]. SSRIs are the most commonly prescribed antidepressants and the best overall treatments for depression patients. However, therapeutic outcomes of SSRIs are often far from satisfactory for both patients and prescribing physicians [32]. Therefore, after having focused clinical research on the development of new drugs, growing evidence suggests that an improved application of available drug may still bring substantial benefit to patients [33, 34]. Moreover, there is a gap between the available pharmacological knowledge and its utilization in health care. The newest initiative to bridge this gap is “Precision Medicine.” It considers individual variability to build the evidence base needed to guide clinical practice [35]. Therapeutic drug monitoring (TDM) is a patient management tool for precision medicine [36]. It enables tailoring the dosage of the medications to the individual patient by combining the quantification of drug concentration in blood, information on drug properties, and patient characteristics [37]. Because patients differ in their ability to absorb, distribute, metabolize, and excrete drug due to concurrent disease, age, concomitant medication or genetic abnormalities, the drug's steady-state concentration in the body may have a more than 20-fold interindividual variation when the same dose of drug is administrated [38, 39]. TDM quantifies the drug's concentration in plasma or serum to adjust the dosage of individual patients, which increases probability of response and decreases risk of adverse drug reactions/toxicity [40, 41]. Moreover, TDM has the potential to enhance the cost-effectiveness of antidepressant therapy [42–44]. The benefits of TDM for optimization of pharmacotherapy, however, can only be obtained when the method is adequately integrated into the clinical treatment process. Current TDM use in depression care is often suboptimal as demonstrated by systematic studies [45–47]. The suboptimal use of TDM wastes laboratory resources and bears the risk of misleading results that will adversely influence clinical decision making. Studies on TDM for antidepressant will further specify the information on the imperfect use of TDM [48].

Among SSRIs, citalopram is the most SSRI [13], and some studies reported that it is more effective and better tolerated than other drugs for depression but has been associated with suicidality and worsening depression especially in adolescents and young adults [49]. Citalopram is strongly recommended for TDM by the Arbeitsgemeinschaft für Neuropsychopharmakologie und Pharmakopsychiatrie (AFNP) guidelines and was recently upgraded into the level 1 recommendation drug [37, 50]. Its reported therapeutic reference ranges (50–110 ng/mL) are established and have been quantified. Controlled clinical trials have known beneficial effects of TDM, reports on decreased tolerability or intoxications [50]. Fluoxetine strongly inhibits 5-HT uptake with minimal effects on other neurotransmitter uptake system [51]. Norfluoxetine, an active metabolite of fluoxetine, contributes to the long elimination half-life (3-15 days) and overall clinical effect of fluoxetine [12]. TDM of fluoxetine is listed as "useful" AFNP guidelines [37, 50]. The therapeutic reference range of 120–500 ng/mL includes the quantification of fluoxetine and its long-lasting active metabolite, norfluoxetine. The total concentration of fluoxetine and norfluoxetine in plasma is needed to be determined. Thus, there is a clinical demand for the detection of fluoxetine and norfluoxetine when patients are receiving fluoxetine. The clinical service for TDM of antidepressants needs to be established.

Acknowledgements

The authors gratefully acknowledge Victoria Li, Xiao Li, and Curt Becker for proofreading the draft of this chapter.

Author details

Ying Qu
Leulan Bioscience, San Diego, California, USA

*Address all correspondence to: yingqu68@gmail.com

References

[1] Feldman R, Meyer J, Quenzer L. Principles of Neuropsychopharmacology. Sunderland, MA, USA: Sinauer Associates, Inc.; 1997

[2] Berger M, Gray JA, Roth BL. The expanded biology of serotonin. Annual Review of Medicine. 2009;**60**:355-366

[3] Jakobsen JC, Katakam KK, Schou A, Hellmuth SG, Stallknecht SE, Leth-Moller K, et al. Selective serotonin reuptake inhibitors versus placebo in patients with major depressive disorder. A systematic review with meta-analysis and trial sequential analysis. BMC Psychiatry. 2017;**17**:58

[4] Fournier JC, DeRubeis RJ, Hollon SD, Dimidjian S, Amsterdam JD, Shelton RC, et al. Antidepressant drug effects and depression severity: A patient-level meta-analysis. JAMA. 2010;**303**:47-53

[5] Duda CT, Kissinger PT. Methods in Neurotransmitter and Neuropeptide Research. Amsterdam: Elsevier; 1993

[6] Qu Y, Moons L, Vandesande F. Determination of serotonin, catecholamines and their metabolites by direct injection of supernatants from chicken brain tissue homogenate using liquid chromatography with electrochemical detection. Journal of Chromatography. B, Biomedical Sciences and Applications. 1997;**704**:351-358

[7] Qu Y, Vandesande F, Arckens L. Identification and quantification of monoaminergic neuromodulators in the sub-cortical region of cat visual cortex by microbore HPLC-ED and protein assay. Journal of Liquid Chromatography and Related Technologies. 2001;**24**:2087-2100

[8] Qu Y, Eysel UT, Vandesande F, Arckens L. Effect of partial sensory deprivation on monoaminergic neuromodulators in striate cortex of adult cat. Neuroscience. 2000;**101**:863-868

[9] Barbier AJ, Aluisio L, Lord B, Qu Y, Wilson SJ, Boggs JD, et al. Pharmacological characterization of JNJ-28583867, a histamine H(3) receptor antagonist and serotonin reuptake inhibitor. European Journal of Pharmacology. 2007;**576**:43-54

[10] Fuller RW. Serotonin uptake inhibitors: Uses in clinical therapy and in laboratory research. Progress in Drug Research. 1995;**45**:167-204

[11] Koch S, Perry KW, Nelson DL, Conway RG, Threlkeld PG, Bymaster FP. R-fluoxetine increases extracellular DA, NE, as well as 5-HT in rat prefrontal cortex and hypothalamus: An in vivo microdialysis and receptor binding study. Neuropsychopharmacology. 2002;**27**:949-959

[12] Qu Y, Aluisio L, Lord B, Boggs J, Hoey K, Mazur C, et al. Pharmacokinetics and pharmacodynamics of norfluoxetine in rats: Increasing extracellular serotonin level in the frontal cortex. Pharmacology, Biochemistry, and Behavior. 2009;**92**:469-473

[13] Qu Y, Olson L, Jiang XH, Aluisio L, King C, Jones EB, et al. Evaluating PK/PD relationship of CNS drug by using liquid chromatography/tandem mass spectrometry coupled to in vivo microdialysis. In: Prasain JK, editor. Tandem Mass Spectrometry - Applications and Principles. London, UK: IntechOpen; 2012. pp. 421-440

[14] World Health Organization. Depression key fact. 2018. Available from: http://www.who.int/news-room/fact-sheets/detail/depression

[15] Menza MA, Kaufman KR, Castellanos A. Modafinil augmentation

of antidepressant treatment in depression. The Journal of Clinical Psychiatry. 2000;**61**:378-381

[16] Keith JM, Gomez LA, Letavic MA, Ly KS, Jablonowski JA, Seierstad M, et al. Dual serotonin transporter/histamine H3 ligands: Optimization of the H3 pharmacophore. Bioorganic & Medicinal Chemistry Letters. 2007;**17**:702-706

[17] Letavic MA, Keith JM, Jablonowski JA, Stocking EM, Gomez LA, Ly KS, et al. Novel tetrahydroisoquinolines are histamine H3 antagonists and serotonin reuptake inhibitors. Bioorganic & Medicinal Chemistry Letters. 2007;**17**:1047-1051

[18] Letavic MA, Keith JM, Ly KS, Barbier AJ, Boggs JD, Wilson SJ, et al. Novel naphthyridines are histamine H3 antagonists and serotonin reuptake transporter inhibitors. Bioorganic & Medicinal Chemistry Letters. 2007;**17**:2566-2569

[19] Boyer EW, Shannon M. The serotonin syndrome. The New England Journal of Medicine. 2005;**352**:1112-1120

[20] Isbister GK, Bowe SJ, Dawson A, Whyte IM. Relative toxicity of selective serotonin reuptake inhibitors (SSRIs) in overdose. Journal of Toxicology. Clinical Toxicology. 2004;**42**:277-285

[21] Domino FJ, Baldor RA. The 5-Minute Clinical Consult 2014. Philadelphia, US: Lippincott Williams & Wilkins; 2013

[22] Nisijima K, Yoshino T, Yui K, Katoh S. Potent serotonin (5-HT)(2A) receptor antagonists completely prevent the development of hyperthermia in an animal model of the 5-HT syndrome. Brain Research. 2001;**890**:23-31

[23] Qu Y, Chang L, Klaff J, Seeman R, Balbo A, Rapoport SI. Imaging of brain serotonergic neurotransmission involving phospholipase A2 activation and arachidonic acid release in unanesthetized rats. Brain Research. Brain Research Protocols. 2003;**12**:16-25

[24] Wettstein JG, Host M, Hitchcock JM. Selectivity of action of typical and atypical anti-psychotic drugs as antagonists of the behavioral effects of 1-[2,5-dimethoxy-4-iodophenyl]-2-aminopropane (DOI). Progress in Neuro-Psychopharmacology & Biological Psychiatry. 1999;**23**:533-544

[25] Qu Y, Chang L, Klaff J, Balbo A, Rapoport SI. Imaging brain phospholipase A2 activation in awake rats in response to the 5-HT2A/2C agonist (+/−)2,5-dimethoxy-4-iodophenyl-2-aminopropane (DOI). Neuropsychopharmacology. 2003;**28**:244-252

[26] Qu Y, Chang L, Klaff J, Seemann R, Rapoport SI. Imaging brain phospholipase A2-mediated signal transduction in response to acute fluoxetine administration in unanesthetized rats. Neuropsychopharmacology. 2003;**28**:1219-1226

[27] Blier P, de Montigny C. Current advances and trends in the treatment of depression. Trends in Pharmacological Sciences. 1994;**15**:220-226

[28] Qu Y, Chang L, Klaff J, Seemann R, Greenstein D, Rapoport SI. Chronic fluoxetine upregulates arachidonic acid incorporation into the brain of unanesthetized rats. European Neuropsychopharmacology. 2006;**16**:561-571

[29] Qu Y, Villacreses N, Murphy DL, Rapoport SI. 5-HT2A/2C receptor signaling via phospholipase A2 and arachidonic acid is attenuated in mice lacking the serotonin reuptake transporter. Psychopharmacology. 2005;**180**:12-20

[30] Giovacchini G, Chang MC, Channing MA, Toczek M, Mason A,

Bokde AL, et al. Brain incorporation of [11C]arachidonic acid in young healthy humans measured with positron emission tomography. Journal of Cerebral Blood Flow and Metabolism. 2002;**22**:1453-1462

[31] Bergfeld IO, Mantione M, Figee M, Schuurman PR, Lok A, Denys D. Treatment-resistant depression and suicidality. Journal of Affective Disorders. 2018;**235**:362-367

[32] Adli M, Baethge C, Heinz A, Langlitz N, Bauer M. Is dose escalation of antidepressants a rational strategy after a medium-dose treatment has failed? A systematic review. European Archives of Psychiatry and Clinical Neuroscience. 2005;**255**:387-400

[33] Ceskova E. The need to improve current psychopharmacotherapy before developing new drugs. Expert Opinion on Pharmacotherapy. 2014;**15**:1969-1973

[34] Shin C, Han C, Pae CU, Patkar AA. Precision medicine for psychopharmacology: A general introduction. Expert Review of Neurotherapeutics. 2016;**16**:831-839

[35] Collins FS, Varmus H. A new initiative on precision medicine. The New England Journal of Medicine. 2015;**372**:793-795

[36] Jang SH, Yan Z, Lazor JA. Therapeutic drug monitoring: A patient management tool for precision medicine. Clinical Pharmacology and Therapeutics. 2016;**99**:148-150

[37] Hiemke C, Bergemann N, Clement HW, Conca A, Deckert J, Domschke K, et al. Consensus guidelines for therapeutic drug monitoring in neuropsychopharmacology: Update 2017. Pharmacopsychiatry. 2018;**51**:9-62

[38] Egberts KM, Mehler-Wex C, Gerlach M. Therapeutic drug monitoring in child and adolescent psychiatry. Pharmacopsychiatry. 2011;**44**:249-253

[39] Hermann M, Waade RB, Molden E. Therapeutic drug monitoring of selective serotonin reuptake inhibitors in elderly patients. Therapeutic Drug Monitoring. 2015;**37**:546-549

[40] Qu Y, Brady K, Apilado R, O'Malley T, Reddy S, Chitkara P, et al. Capillary blood collected on volumetric absorptive microsampling (VAMS) device for monitoring hydroxychloroquine in rheumatoid arthritis patients. Journal of Pharmaceutical and Biomedical Analysis. 2017;**140**:334-341

[41] Qu Y, Noe G, Breaud AR, Vidal M, Clarke WA, Zahr N, et al. Development and validation of a clinical HPLC method for the quantification of hydroxychloroquine and its metabolites in whole blood. Future Science OA. 2015;**1**:FSO26

[42] Ostad Haji E, Mann K, Dragicevic A, Muller MJ, Boland K, Rao ML, et al. Potential cost-effectiveness of therapeutic drug monitoring for depressed patients treated with citalopram. Therapeutic Drug Monitoring. 2013;**35**:396-401

[43] Akerblad AC, Bengtsson F, von Knorring L, Ekselius L. Response, remission and relapse in relation to adherence in primary care treatment of depression: A 2-year outcome study. International Clinical Psychopharmacology. 2006;**21**:117-124

[44] von Knorring L, Akerblad AC, Bengtsson F, Carlsson A, Ekselius L. Cost of depression: Effect of adherence and treatment response. European Psychiatry. 2006;**21**:349-354

[45] Loayza N, Crettol S, Riquier F, Eap CB. Adherence to antidepressant

treatment: What the doctor thinks and what the patient says. Pharmacopsychiatry. 2012;**45**:204-207

[46] Sharma S, Joshi S, Mukherji S, Bala K, Tripathi CB. Therapeutic drug monitoring: Appropriateness and clinical utility in neuropsychiatry practice. American Journal of Therapeutics. 2009;**16**:11-16

[47] Vuille F, Amey M, Baumann P. Use of plasma level monitoring of antidepressants in clinical practice. Towards an analysis of clinical utility. Pharmacopsychiatry. 1991;**24**:190-195

[48] Mann K, Hiemke C, Schmidt LG, Bates DW. Appropriateness of therapeutic drug monitoring for antidepressants in routine psychiatric inpatient care. Therapeutic Drug Monitoring. 2006;**28**:83-88

[49] Sharbaf Shoar N, Padhy RK. Citalopram. Treasure Island (FL): StatPearls; 2018

[50] Schoretsanitis G, Paulzen M, Unterecker S, Schwarz M, Conca A, Zernig G, et al. TDM in psychiatry and neurology: A comprehensive summary of the consensus guidelines for therapeutic drug monitoring in neuropsychopharmacology, update 2017. The World Journal of Biological Psychiatry. 2018;**19**:162-174

[51] Beasley CM, Masica DN, Potvin JH. Fluoxetine: A review of receptor and functional effects and their clinical implications. Psychopharmacology. 1992;**107**:1-10

Chapter 2

Serotonin Reuptake Inhibitors and Their Role in Chronic Pain Management

Adela Hilda Onuțu, Dan Sebastian Dîrzu and Cristina Petrișor

Abstract

Serotonin has a particular place in the modulation of pain. Experimental studies have described 5-HT1–7 receptors and their effects on facilitation or inhibition of nociceptive input. Selective serotonin reuptake inhibitors (SSRIs) and serotonin-norepinephrine reuptake inhibitors showed efficient and safer than tricyclic antidepressants in neuropathic pain. Although there is evidence regarding the beneficial impact of SSRIs in the multimodal acute pain management, studies are still searching for the potentially favorable effect of these drugs in the prevention of chronic postoperative pain. The scope of this chapter would be to update the knowledge regarding serotonin involving in pain pathways and to highlight the importance and contribution of serotonin reuptake inhibitors in the multimodal pain management schemes.

Keywords: serotonin, pain, SSRIs, SNRIs, pain management

1. Introduction

Chronic pain is recognized today as a disease [1], affects almost 20% of the population [2], and represents a significant cause of disability bringing along high secondary social costs. The management of chronic pain involves pharmacological and interventional tools and become a priority for healthcare systems. This chapter aims to summarize the role of serotonin reuptake inhibitors (SRI) in the treatment of chronic pain. SRI includes selective serotonin reuptake inhibitors (SSRI) and serotonin-norepinephrine reuptake inhibitors (SNRI).

2. Chronic pain conditions

Defined as the pain that lasts more than 3 months, frequent after the disappearance of the causal factor, chronic pain shows numerous risk factors (socio-demographic, biological, clinical, and psychological). Thus, most affected are females, older people, and people with low socio-economic level. A significant risk for developing chronic pain is the pain itself (acute and chronic at other sites), and also are incriminated the geographical background, occupational factors, and the history of abuse and violence. Neuroimaging studies have already proved the changes in the

IntechOpen

brain with severe pain, reversible with proper treatment, and also suggested its importance for preventing the chronicization of pain [3]. The treatment of chronic pain is multi/interdisciplinary and multimodal, targeting different mechanisms of pain. We summarized some critical pain syndromes, which benefit from the SRI medication.

2.1 Diabetic neuropathy

Diabetes mellitus affects billions of people worldwide. The painful diabetic neuropathy (PDPN) occurs in 20% of diabetes patients during the disease. Risk factors include age, hypertension, obesity, alcohol abuse, and smoking.

Pathogenesis implies endoneurial microangiopathy and axonal loss, especially in sensory nerves. Aldose reductase activation by increasing polyol flux and the deposition of advanced glycated end-products are the primary determinants of PDPN. Secondary ischemia leads to enhanced oxidative stress and high production of free radicals, which leads to nerve damage [4].

Clinical PDPN may present as burning, stabbing, dull and aching, or sharp pain. In some instances, allodynia (painful response to a normally non-noxious stimulus) might accompany pain. PDPN is symptomatic mainly in the lower limbs and progresses proximally. Patients with PDPN show skin changes and loss of sensory that could lead further to diabetic ulcers.

The medication of painful diabetic neuropathy includes duloxetine, venlafaxine, tricyclic antidepressants (TAD), oxcarbazepine, and tapentadol. Overall, the quality of life in patients with PDPN is poor [5].

2.2 Fibromyalgia

Fibromyalgia (FM) is a syndrome composed of widespread chronic pain, muscle fatigue, and functional symptoms. It shows a genetic predisposition, but environmental factors play a prominent place during the disease. FM pathogenesis involves modified inflammatory response and oxidative stress [6].

Diagnosis is difficult because of the variety of clinical symptoms—75% of these patients do not meet the inclusion criteria, thus often they lack the diagnosis. Besides, these patients develop sleep disturbances and sexual dysfunction, altering further the quality of life.

The current evidence suggest for FM management antidepressants, cardiovascular exercise, and cognitive behavioral therapy [7]. Meta-analysis results agree that the medication approved by FDA—milnacipran—and duloxetine are effective in FM while there are concerns that the results showed only a moderate effect on pain and sleep, and no impact on fatigue [8].

2.3 Tension-type headache

Tension-type headache (TTH) is a typical headache (up to 78%), caused by the contractions of muscles of the scalp, neck, and jaw, and triggered mainly by stress and emotional conflicts. It is described as a moderate pressure applied to the frontal area, around the head or neck, and according to its frequency is classified as infrequent, frequent, and chronic.

Chronic TTH results as a consequence of sensitization of the pain pathway due to persistent pericranial myofascial nociceptive input. This TTH shows a frequency of at least 15 days/month for at least 3 months. If nausea and vomiting are present, exclude the diagnosis of TTH. Photophobia may occur in TTH.

Treatment of the acute episodes of TTH includes nonsteroidal anti-inflammatory drugs and acetaminophen, while their prevention associates pharmacologic and non-pharmacologic (physical and psychologic therapy) interventions. Tricyclic antidepressants (amitriptyline) are the most studied drugs in TTH, but new studies showed efficacy for other antidepressants including SRI—citalopram, sertraline, venlafaxine, and paroxetine [9].

2.4 Somatoform pain

Somatoform pain (SP) is the primary symptom in an ambiguous and unclarified category called somatization spectrum disorders (SSD), defined as the displaying of somatic complaints as a result of social stress. It shows a growing incidence (up to 60%) and is a symptom generally unexplained by the medical condition of these patients (which must be ruled out). The symptomatology—headache and musculoskeletal pain—overlaps with other chronic pain syndromes and may be associated with psychiatric symptoms (depression, anxiety, personality disorders) and thus makes the diagnosis difficult. The mechanism of this condition is a subject of debate, but a genetic predisposition plus an altered interpersonal relationship in childhood and adolescence are the determining factors [10].

Treatment is focused on psychotherapy and modulation of interpersonal relations, by learning to develop robust, safe, and supportive social relationships. Besides, acupuncture and massage proved efficacy. Medication includes TADs and SSRIs [11].

3. Selective serotonin reuptake inhibitors in chronic pain management

Due to the association between chronic pain states and depression and also due to the continuous need and search for effective analgesic drugs, antidepressants have long been considered for the treatment of chronic pain. Some antidepressants are useful in the management of pain syndromes showing analgesic effects, but not all antidepressants have analgesic properties [12]. TADs are recognized to have analgesic effects in doses lower than the antidepressant ones. However, frequent side effects preclude their widespread use.

Consequently, newer generations of antidepressants, like SSRIs and SNRIs, have been studied in chronic pain management. For SSRIs, efficiency in chronic pain conditions has been debated, and results are still inconclusive. It is felt that antidepressants with both noradrenergic and serotoninergic activities are superior analgesics compared to drugs that possess only serotoninergic activity [13].

Currently available SSRIs are fluoxetine, sertraline, paroxetine, citalopram, and escitalopram. Fluvoxamine is approved for the treatment of obsessive-compulsive disorders but has sometimes been used off-label for the treatment of depression. SSRIs are currently approved and used for the treatment of a wide range of diseases: depression, anxiety and panic disorders, obsessive-compulsive disorders, post-traumatic stress disorder, premenstrual dysphoric syndrome, dysthymia, irritable bowel syndrome, eating disorders, alcohol abuse, and some personality disorders [14].

SSRIs utility for the treatment of chronic pain has been questioned but seems attractive due to their better side-effect profile compared to first-generation antidepressants like TADs, as SSRIs selectively block serotonin reuptake (reabsorption in the synaptic cleft).

How could SSRIs be useful in chronic pain management? Do they possess both antidepressant and intrinsic analgesic properties?

Even though widely prescribed, the mechanism of action of SSRIs is not yet fully understood. The traditional theories claim the fact that antidepressant drugs act by influencing certain brain neurotransmitters [15]. Serotonin (5-HT) is one of the neurotransmitters which carry signals between neurons. The monoamine signaling theory of depression explains how SSRIs and other antidepressants work at the synaptic level by inhibiting the reuptake of one or several neurotransmitters, an effect which is almost immediate and leads to the increase of the extracellular level of the mediator. SSRIs are selective inhibitors of the presynaptic 5-HT reuptake transporter (SERT) that leads to an acute increase in serotonin concentrations in the synaptic cleft. This effect does not explain why antidepressant drugs work 2–4 weeks after treatment commencement, which might be better explained by receptor downregulation and delayed desensitization of presynaptic serotonin receptors [16]. Recent findings also suggest changes in brain-derived neurotrophic factor expression, which might even lead to SSRI antidepressant effect. Another newer theory suggests that SSRIs impact brain levels of allopregnanolone, enhancing gamma-amino butyric functions in the brain [14]. Apart from monoamine neurotransmitter's imbalance, the inflammatory theory of depression claims the increased serum levels of proinflammatory mediators in the depressed patients [17]. As inflammation is the well-known cause of acute and some type of chronic pain, proinflammatory mediators play the capital role in initiating nociception and peripheral sensitization. *In vitro* experimental studies and early *in vivo* studies suggested that SSRIs could inhibit the release of TNF-α, interferon γ, interleukin 1β, and free radical superoxide [16, 18]. Probably one of the most plausible humoral links between chronic pain conditions and depression is inflammation. If SSRIs have intrinsic anti-inflammatory and anti-oxidative properties and could modulate inflammatory processes, then this could be an explanation for their therapeutic effect in chronic pain management. The detailed specificity of action for this mechanism remains unknown [19]. Intrinsic antihyperalgesic effects in animal models have recently been described for SSRIs [20–23].

Possible side effects observed during antidepressant treatment with SSRIs also need to be considered when prescribing SSRIs for chronic pain management.

These side effects include [12, 14, 24]:

- Drowsiness, dry mouth, blurred vision, dizziness
- Gastrointestinal effects: nausea, diarrhea or constipation, vomiting
- Central nervous system effects: insomnia, agitation or restlessness, headache, tremors, increased sweating, rarely extrapyramidal symptoms, anorexia
- Syndrome of inappropriate antidiuretic hormone secretion with hyponatremia, somnolence, delirium, confusion
- Sexual dysfunction
- Weight gain
- Platelet dysfunction and increased risk of bleeding
- Drug interactions due to the concomitant hepatic metabolism involving the cytochrome P450
- Safety issues in pregnancy

- Serotonin syndrome

Suicide might be a risk occurring early in the treatment, even though larger epidemiological studies do not confirm this assumption [14].

SSRIs discontinuation syndrome is characterized by sensory and gastrointestinal symptoms, dizziness, lethargy, and sleep disturbances [25].

3.1 Individual SSRIs and their efficiency in chronic pain conditions as highlighted in clinical trials

3.1.1 Fluoxetine

Fluoxetine (Prozac™, Sarafem™) has been one of the first SSRIs available for the treatment of depression. Its use for chronic pain management has been highlighted in several clinical trials including modest numbers of patients (**Table 1**). For chronic tension-type headache, fluoxetine administered in 20 mg daily dose is equally efficient to desipramine [26]. For the treatment of painful diabetic neuropathy, fluoxetine is no more effective than placebo and ameliorates pain in 48% of the patients, especially the depressed ones [27]. For somatoform pain disorders, the analgesic effect is related to treatment duration and is related to its antidepressant effect as depressive patients show greater improvement compared to non-depressed ones [28]. Fluoxetine was found to be efficient for the treatment of fibromyalgia when compared to placebo or amitriptyline [29, 30].

3.1.2 Fluvoxamine

Fluvoxamine (Luvox™) is currently used for the treatment of obsessive-compulsive disorders, and the therapeutic dose varies widely between 50 and 300 mg. Non-depressed patients with severe chronic tension-type headache respond to fluvoxamine 50–100 mg daily [31], and it is efficient in central post-stroke pain, cancer pain, and osteoarthritis [32–34]. However, for chronic can cer pain, its beneficial effect has not yet been proven [35].

	Study	Chronic pain condition	Dose used for chronic pain (mg)	No patients	Comparator	Efficiency
Fluoxetine 10–80 mg/day for depression [14]	Walker et al. [26]	Chronic tension-type headache	20	25	Desipramine	Equally efficient
	Max et al. [27]	Diabetic neuropathy	40	46	Placebo	Equally efficient
	Luo et al. [28]	Somatoform pain disorders	20	80	Placebo	Efficient for depressed patients
	Goldenberg et al. [29]	Fibromyalgia	20	19	Amitriptyline	Effective
	Arnold et al. [30]	Fibromyalgia	45 ± 25	60	Placebo	Effective

Table 1.
Randomized controlled trials for fluoxetine in chronic pain management.

3.1.3 Sertraline

Sertraline (Zoloft™) is recommended in single daily doses of 50–200 mg for the treatment of depression. In small sample size studies, it has proven to be efficient in non-cardiac chronic chest pain and chronic pelvic pain of prostatic origin in men [36, 37], but not in women with chronic pelvic pain [38].

3.1.4 Paroxetine

Paroxetine (Paxil™, Seroxat™) is one of the most extensively studied SSRI for chronic pain management (**Table 2**). For tension-type daily headache, two studies failed to prove any beneficial effect [39, 40]. Foster et al. suggested that by extending treatment periods up to 3–9 months, patients may benefit [41]. For chronic low back pain, doses of 20 mg are less efficient than maprotiline, and the effects are similar to placebo [42, 43]. In fibromyalgia, paroxetine improves overall symptomatology, but the effect on pain is less robust [44]. Paroxetine has been shown to be useful for the treatment of diabetic peripheral neuropathy, but not more efficient than imipramine [45]. In a mixed study comparing paroxetine and citalopram versus gabapentin, the comparable efficiency of these two SSRIs with gabapentin was shown [46].

	Study	Chronic pain condition	Dose used for chronic pain (mg)	No of patients	Comparator	Efficiency
Paroxetine 10–50 mg daily for depression [14]	Langemark and Olesen [39]	Chronic tension-type headache	20–30	50	Sulpiride	Less efficient
	Holroyd et al. [40]	Chronic headache, non-responding to amitriptyline	Up to 40	31	Placebo	Modest effect
	Foster and Bafaloukos [41]	Chronic daily headache	10–50	48	Placebo	Efficient when used for 3–9 months
	Dickens et al. [42]	Chronic low back pain	20	91	Placebo	Not efficient
	Atkinson et al. [43]	Chronic low back pain	20	74	Maprotiline	Less efficient
	Patkar et al. [44]	Fibromyalgia	12.5–2.5 mg	116	Placebo	Inconclusive
	Sindrup et al. [45]	Diabetic peripheral neuropathy	40	29	Placebo and imipramine	Efficient compared to placebo, less efficient compared to imipramine
	Giannopoulos et al. [52]	Diabetic peripheral neuropathy	20–40	101	Citalopram or paroxetine *versus* gabapentin	Comparable efficiency

Table 2.
Randomized controlled trials for paroxetine in chronic pain management.

	Study	Chronic pain condition	Dose used for chronic pain (mg)	No of patients	Comparator	Efficiency
Citalopram 10–80 mg for depression [14]	Nørregaard et al. [46]	Fibromyalgia	20–40	43	Placebo	No effect
	Anderberg et al. [47]	Fibromyalgia	20–40	40	Placebo	Inconclusive
	Aragona et al. [48]	Somatoform pain disorder	20	35	Reboxetine	Moderate effect
	Bendsten et al. [49]	Chronic tension-type headache	20	40	Placebo and amitriptyline	No significant effect
	Viazis et al. [50]	Gastroesophageal reflux disease	20	63		Efficient when administered with proton pump inhibitors
	Roohafza et al. [51]	Pediatric functional abdominal pain	20	86	Placebo	Effective
	Giannopoulos et al. [52]	Diabetic peripheral neuropathy	20–40	101	Citalopram or paroxetine *versus* gabapentin	Comparable efficiency

Table 3.
Randomized controlled trials for citalopram in chronic pain management.

3.1.5 Citalopram

Citalopram (Celexa™, Cipramil™) is administered in 10–80 mg dose once daily for the treatment of depression (**Table 3**). It has been investigated for the treatment of fibromyalgia and chronic tension-type headache, with no beneficial results [47–49], while for somatoform pain disorders it has only moderate analgesic effect [50–52].

3.1.6 Escitalopram

Escitalopram (Cipralex™, Lexapro™) has antidepressive effects in 10–20 mg daily dose. For chronic low back pain, citalopram has similar results compared to duloxetine [53]. It has pain-relieving effects in painful diabetic neuropathy and somatoform disorders [54, 55]. For the treatment of pain symptoms associated with depression, escitalopram is equally effective with nortriptyline [56].

4. Serotonin norepinephrine reuptake inhibitors in pain management

SNRIs are first-line antidepressants known to inhibit the reuptake of serotonin and norepinephrine almost exclusively by binding to their transporters (SERT and NET). This category includes drugs with very different chemical structure and includes venlafaxine, desvenlafaxine, duloxetine, milnacipran, and levomilnacipran.

SNRIs show different pharmacokinetics and dynamics and also different affinity to SERT and NET with consequences on their therapeutic actions (**Table 4**).

SNRI	Bioavailability (%)	Elimination half life	Elimination	SERT affinity	NAT affinity	Active metabolite
Venlafaxine	45	5 h (IR) 11 h (ER)	Renal	High	Low	Yes
Duloxetine IR	50	12 h	Renal + feces	High	High	No
Milnacipran	85	8 h	Renal (55% unchanged)	Moderate	Moderate	No
Desvenlafaxine	80	11 h (IR) 13–14 h (ER)	Renal (45% unchanged) at 72 h	High	Low	No
Levomilnacipran	92	12 h	Renal 58% unchanged	Low	High	No

SERT, serotonin transporter; NAT, noradrenaline transporter; IR, immediate release; ER, extended release.

Table 4.
SNRIs pharmacokinetics and pharmacodynamics.

Side effects of SNRIs are common to all antidepressants, but these drugs add dry mouth and constipation due to increased levels of noradrenaline. The risk of withdrawal because of side effects, in patients with chronic pain, was highest for milnacipran and followed by venlafaxine and duloxetine [57].

4.1 Venlafaxine

Venlafaxine (Effexor™) is an SNRI with mixed action on amine reuptake. When administrated in low doses, it inhibits SERT and at higher doses NAT. It is indicated for major depressive disorder (MDD) and also for anxiety, panic disorders, and social phobia management.

An experimental study showed its antihyperalgesic effect after a single administration in a diabetic neuropathic pain model, a result reversed by pretreatment with yohimbine and chloroamphetamine, but not by naloxone [58].

Long ago, a short case report raised attention to the potential beneficial effect of venlafaxine in chronic pain management [59], and later others confirmed its beneficial effects in managing neuropathic pain: peripheral neuropathy, postherpetic neuropathy, headache, and multiple sclerosis. In a systematic review on neuropathic pain, the authors found four trials (high quality evidence): two with positive results at doses of 150–225 mg venlafaxine ER daily and two with negative results (lower doses). The number needed to treat (NNT) was 6.4 and the number needed to harm (NNH) was 11.8 [60].

In elderly patients with low back pain and depression, 150 mg venlafaxine showed efficacy, but the authors suggested that patients who did not respond to small doses may benefit from dose augmentation after a 2-week period [61].

Venlafaxine may be useful in the treatment of spinal cord injury (SCI) associated with MDD because this medication improved SCI-related disability and pain. Still, further trials are needed to determine optimal doses and efficiency in patients with SCI without MDD [62, 63].

Studies in patients with taxane-oxaliplatin-induced neurotoxicity showed clinical improvement after venlafaxine (37.5 mg bid) [64], and further studies are in progress [65].

Venlafaxine had good results in acute pain; in patients with cancer breast surgery, the preoperative administration of 37.5 mg venlafaxine reduced the postoperative opioid consumption and the incidence of chronic postoperative pain at 6 months [66].

A former Cochrane meta-analysis reported little evidence to support the recommendation of venlafaxine in neuropathic pain management and noted that venlafaxine promoted fatigue, nausea, dizziness, and somnolence with a low incidence [67].

Eventually, two recent reviews (11 and 13 trials) found that venlafaxine was beneficial in neuropathic pain management with good tolerability claiming the necessity for further research to expand these findings [68, 69]. There are contradictory findings in these recent reviews, but there is need for further good quality evidence.

4.2 Desvenlafaxine

Desvenlafaxine (Pristiq™) is the third SNRI with FDA approval and only indication for MDD management (50–400 mg daily). The daily recommended dose is 50 mg. Desvenlafaxine is the salt of an active metabolite of venlafaxine, and the ER form allows 1 day administration. It presents a good bioavailability (**Table 4**) and shows a low binding to plasma proteins (30%). Desvenlafaxine binds to SERT 10 times more than to NAT and also has a weak affinity for dopamine transporter. Adverse effects are dose-dependent and typical to all antidepressants. Doses of 200–400 mg showed efficacy in DPN management, with effect sizes similar to duloxetine [70] and with increased side effects at higher doses. At the moment, there is a lack of evidence to support the use of desvenlafaxine in chronic pain management.

4.3 Duloxetine

Duloxetine (Cymbalta™; DLX) is probably the most used drug from this class of antidepressants. Aside from MDD and urinary incontinence, duloxetine is indicated for anxiety disorder, chronic pain in diabetic neuropathy, fibromyalgia, musculoskeletal pain, and osteoarthritis. DLX is a potent SNRI, with a high affinity for both SERT and NAT. It has a moderate bioavailability, with an elimination half-time of 12 h. It is metabolized in the liver and does not possess any active metabolite. Duloxetine exerts antihyperalgesic and allodynic effects, by impairing nociception at a peripheral level (blocks NaV 1.7 current) and by inhibiting neuronal firing [71]. With acute administration, DLX leads to elevated levels of NA and 5-HT, and with chronic treatment, it does not affect further basal levels of these monoamines [72]. Even if the significant pain-relieving effect was found after 7 weeks of treatment [73], others showed that patients treated with DLX for OA knee pain or low lumbar pain who have <10% reduction in pain after 4 weeks treatment have low chance to reach moderate pain reduction by the end of 12 weeks [74]. DLX's recommended dose for the first week is 30 mg, raising the dose to 60 mg in the second week in order to avoid a high incidence of side effects.

Because of interfering with platelet function, it is indicated to stop its administration 4 days before surgery.

Data from animal pain models and clinical studies on DLX administration in perioperative setting (spine, knee, breast surgery) suggested its analgesic effects. Pre- and postoperative duloxetine reduced 24-h opioid consumption, delayed first analgesic requirement, and reduced incidence of chronic postoperative pain at 6 months, being of primary interest for patients with preoperative chronic pain and spine surgery [75–77]; results from ongoing studies will respond to questions remained unanswered. Duloxetine shows good tolerability with dizziness and nausea, dry mouth, and constipation, as more frequent side effects [78].

Study	Pathology	No trials	Number of patients	Findings
Lunn et al. [79] 60 mg	Diabetes fibromyalgia	14 8-DPN; 6-FM	6407	In both category showed efficacy DPN-NNT 5 FM-NNT 8
Quilici et al. [80]	Diabetes	11	679	Effective in DPN NNT = 5 Good toleration Superior to placebo Discontinuation due to AE
Wang et al. [81] 60/120 mg (QD)	Knee osteoarthritis	3	1011	Significant pain reduction Improved function Reported "acceptable" AE
Lee and Song [82]	Fibromyalgia	9	5140	Results showed equal efficacy and tolerability
Hauser et al. [83]	Fibromyalgia	10	6038	Small benefit over placebo

QD, quaque die; DLX, duloxetine; MLC, milnacipran; DPN, diabetes polyneuropathy; NNT, number needed to treat; FM, fibromyalgia; AE, adverse effect.

Table 5.
Meta-analysis for duloxetine in chronic pain management.

While duloxetine proved its efficacy in chronic nociceptive/neuropathic pain [79–83] (**Table 5**), it is yet unrevealed its possible impact on acute postoperative pain and chronic postoperative pain.

4.4 Milnacipran

Milnacipran (Savella™) described in 1998 by Briley as a potent SNRI that showed similar inhibition on both monoamine re-uptakes, in vitro and in vivo, was approved in Europe for the treatment of depression. It did not link to alpha adrenoreceptors, muscarinic cholinergic, and histaminic receptors and showed no effect on beta-adrenergic receptors sensitivity, thus having reduced side effects. The drug has an excellent bioavailability with a mean peak plasma concentration reached between 0.5 and 4 h after the oral administration. About 13% binds to plasma proteins and is wholly eliminated after 36 h [84]. Studies on the efficacy of milnacipran in psychiatric patients revealed its significant superiority when compared to SSRIs. Most frequent adverse effects were nausea, dry mouth, and headaches [85]. Milnacipran has FDA approval for the management of fibromyalgia.

In 2006, Obata et al. found that intrathecal administration of milnacipran reduced allodynia in a rat neuropathic pain model [86].

Other experimental data confirmed these findings regarding milnacipran's antiallodynic and antihyperalgesic effects [87] and showed its effectiveness in treating allodynia in vincristine-induced neuropathic pain [88].

In a Cochrane meta-analysis, Cording et al. analyzed six studies (4238 patients) that compared milnacipran 100/200 mg with placebo in fibromyalgia. By using a "conservative" method of analysis, they found 26% positive response with milnacipran as compared to 19% for placebo and an increased rate of side effects [89].

Despite the evidence that milnacipran (100 or 200 mg) was found to be useful in neuropathic pain, as compared with placebo, Derry et al.'s meta-analysis did not obtain enough data to confirm former data and support its recommendation in chronic neuropathic pain [90]. Future trials are needed to establish milnacipran's possible favorable effects in pain management.

4.5 Levomilnacipran

Levomilnacipran (Fetzima™) is the enantiomer of milnacipran with the highest activity, and its primary indication is MDD. At usual doses, this drug is known to possess a higher potency for norepinephrine (twofold) reuptake inhibition, as compared with 5-HT [91]; but with higher doses, it showed equal efficacy in increasing 5-HT and NE levels [92].

Regarding tolerability, the most frequently recorded adverse effects were nausea, constipation, and sweating, although a small proportion (3–6%) of patients recorded increased blood pressure and heart rate [93]. We have not found any data regarding its use in chronic pain patients.

5. Double function serotonin reuptake inhibitors

A particular category of drugs includes SRIs with double mechanism: 5-HT reuptake inhibition and interaction with 5-HT receptors. Animal studies have suggested that these receptors are included in the descending pain inhibitory systems [94, 95], and their activation is involved in reducing the acute nociceptive and neuropathic pain [96].

5.1 Trazodone

Trazodone (Desyrel™, Oleptro™) is the first non-tricyclic antidepressant approved for the treatment of MDD (1981), and it is also used to treat anxiety, alcohol dependence, insomnia, and chronic pain (off-label). It was developed for the treatment of "mental pain," which was recognized to occur in depression [97]. It acts as a SRI, antagonist of 5-HT_{A2} receptor, and a partial agonist for 5-HT_{A1} receptors. Secondary acts as an antagonist to α_1-adrenergic receptors and lacks any effect on cholinergic receptors. The drug shows a 65% oral bioavailability, 90% plasma protein binding capacity, and is metabolized in the liver (via CYP3A4) to an active metabolite—mCPP. The main excretion route is renal, and the biological half-time is 7 h. Side effects are not only shared with the other antidepressants but also list dry mouth, orthostatic hypotension, cardiac arrhythmias, and priapism.

Trazodone showed some efficacy in several chronic pain conditions represented in **Table 6**, but future studies are needed.

5.2 Nefazodone

Nefazodone (Serzone™) is related to trazodone but with fewer side effects. Doses of 300–600 mg are indicated for the treatment of MDD, panic disorders, and aggressive behavior. It acts as an antagonist of 5-HT_{A2} and 5-HT_{C2} receptors and serotonin, norepinephrine, and dopamine reuptake inhibitor. Its effects on the mentioned receptors enhance neurotransmission by an increased binding on the 5-HT_{A1} receptors. Nefazodone shows an affinity for α_1 and less for β-adrenoreceptors and does not interact with muscarinic cholinergic receptors. It has low bioavailability; it is metabolized in the liver (CYP3A4) and has four metabolites (mCPP active). Nefazodone has a biological half-time between 2 and 4 h and is excreted in urine. Frequent side effects are dry mouth, dizziness, and sleepiness, and rare, severe liver damage [98].

Murine studies yielded the capacity of nefazodone to potentiate opioid analgesia by acting through μ_1 and μ_2 receptors without affecting mortality [99]. Other results indicated that rats treated with nefazodone have shown an increased expression of μ-opioid receptors in the area of the central nervous system related to pain perception and modulation [100].

Even if it shows an excellent clinical profile, at this time we found only a two-center open-label study on the efficacy of nefazodone on preventing chronic daily headache. The study included 52 patients who received nefazodone between 100 and 450 mg (300 mg median) for 12 weeks. The results showed significantly lower incidence and intensity of daily headache and a good tolerance for nefazodone [101].

5.3 Vilazodone

Vilazodone (Viibryd™) approved by FDA (2011) for the treatment of MDD is a partial agonist to the 5-HT_{A1} receptor, GABA agonist, and SRI. Currently is presumed that it increases serotoninergic neurotransmission and it shows fast onset and good effect at daily doses between 10 and 40 mg. Vilazodone has 72% bioavailability when it is taken with food, is metabolized in the liver (via CYP3A4), and it did not possess active metabolites. It is excreted in urine and feces and has a biological half-time of 25 h [102]. Side effects include nausea, vomiting, diarrhea, and insomnia (>5%). Sexual adverse reactions and low influence on weight-gain were reported [103]. Even if it presumed that vilazodone should add value in the treatment of patients with the depression-pain syndrome, there are not yet available data on its efficacy in pain states.

Study	Chronic pain condition	Dose	Number of patients	Comparator	Efficiency
Wilson [103]	Diabetic neuropathy	50–100 mg	31	—	Effective
Ventafridda et al. [104]	Deafferentation pain	—	45	Amitriptyline	Equal efficacy
Goodkin et al. [105]	Chronic low back pain	201 mg (average)	42	Placebo	Similar effect
Morillas-Arques et al. [106]	Fibromyalgia	50–300 mg	66	—	Effective
Calandre et al. [107]	Fibromyalgia	50–300 mg trazodone + 75–450 mg pregabalin	41	—	Pregabalin enhanced the favorable effects of trazodone
Davidoff et al. [108]	Dysesthetic pain following spinal cord injury	150 mg	18	placebo	Similar effect
Battistella et al. [109]	Migraine (pediatric 7–18 years)	1 mg/kg	40	placebo	Effective
Frank et al. [110]	Rheumatoid arthritis	1.5 mg/kg	47	—	No effect

Table 6.
Trials for trazodone in chronic pain management.

6. Conclusions

SSRIs seem to be effective in most chronic pain conditions, and they are well tolerated [41]. The efficacy of SSRIs might be comparable to TADs and SNRIs, but their tolerability and safety are superior [30]. For some chronic pain conditions, valuable, while for others their utility is limited:

- For migraine, SSRIs are not better than placebo for reducing the number of attacks, and results of studies on migraine are conflicting [24, 111].
- Patients with chronic tension-type headache seem to benefit from SSRIs [24].
- There are conflicting results regarding the use of SSRIs for pelvic pain.
- Non-cardiac chest pain might benefit from SSRIs.
- Low back pain does not seem to respond well to SSRIs.
- The effects of SSRIs on fibromyalgia are uncertain [112].
- Diabetic neuropathy looks to improve from SSRIs treatment.
- Post stroke central pain might improve with fluvoxamine.
- Evidences support that antidepressants are useful for the treatment of irritable bowel syndrome [113].
- There is no evidence from randomized controlled trials to recommend antidepressants to treat chronic non-cancer pain in children and adolescents [114] or adults.

Even though several clinical trials were published, the results remain inconclusive. That happens because the sample sizes are quite modest rendering the studies slightly underpowered. Primary outcomes are variable: self-reported pain scores, effect on pain symptoms observed by the physician, complex pain questionnaires, and effects on quality of life and functionality. Current drug classes available for chronic pain treatment include anti-inflammatory drugs, opioids, gabapentinoids along with interventional or surgical management, and physical activity. Heterogeneity of the chronic pain syndromes, many currently available drugs and treatment modalities, and drug-drug and drug-interventional management associations should be considered when designing future larger scale trials.

In conclusion, compared to all other antidepressants in the management of chronic pain, for SSRIs, the data are still inconclusive, and studies are fewer in number. For depression, SSRIs are considered first-line agents due to a favorable side-effect profile and good tolerability. However, they have not yet entered first-line use for neuropathic pain conditions [12]. Probably, it would be advisable to restrict their use for those patients failing to respond to other medications or who do not tolerate side effects.

From SNRIs category, in particular duloxetine is already a first-line treatment for DPN and other chronic pain syndromes (fibromyalgia, musculoskeletal pain, and osteoarthritis), showing good results and an acceptable safety profile. It also showed favorable effects on chronic postoperative pain and life quality with the perioperative administration in surgery with a high incidence of chronic pain (spine, breast). Venlafaxine is a drug of choice for the treatment of fibromyalgia. Milnacipran proved antiallodynic and antihyperalgesic effects and might show further positive results in chronic pain management; well-designed trials are still required.

SRIs seem to play their role through the spinal modulation pain pathways being less involved in reducing nociception, and that is probably why their effects are more evident in patients with chronic pain states.

Conflict of interest

Nothing to declare.

Author details

Adela Hilda Onuțu[1*], Dan Sebastian Dîrzu[2] and Cristina Petrișor[1,2]

1 Emergency Clinic County Hospital, Cluj-Napoca, Romania

2 Anaesthesia and Intensive Care II Department, "Iuliu Hatieganu" University of Medicine and Pharmacy, Cluj-Napoca, Romania

*Address all correspondence to: adela_hilda@yahoo.com

References

[1] Tracey I, Bushnell MC. How neuroimaging studies have challenged us to rethink: Is chronic pain a disease? The Journal of Pain. 2009;**10**(11):1113-1120

[2] van Hecke O, Torrance N, Smith BH. Chronic pain epidemiology and its clinical relevance. British Journal of Anaesthesia. 2013;**111**(1):13-18

[3] Bingel U, Tracey I, Wiech K. Neuroimaging as a tool to investigate how cognitive factors influence analgesic drug outcomes. Neuroscience Letters. 2012;**520**(2):149-155

[4] Schreiber AK, Nones CFM, Reis RC, Chichorro J, Cunha JM. Diabetic neuropathic pain: Physiopathology and treatment. World Journal of Diabetes. 2015;**6**(3):432-444

[5] Waldfogel JM, Nesbit SA, Dy SM, Sharma R, Zhang A, Wilson LM, et al. Pharmacotherapy for diabetic peripheral neuropathy pain and quality of life: A systematic review. Neurology. 2017;**88**(20):1958-1967

[6] Staud R. Peripheral pain mechanisms in chronic widespread pain. Best practice and research. Clinical Rheumatology. 2011;**25**(2):155-164

[7] Okifuji A, Gao J, Bokat C, Hare BD. Management of fibromyalgia syndrome in 2016. Pain Management. 2016;**6**(4): 383-400

[8] Macfarlane GJ, Kronisch C, Dean LE, Atzeni F, Häuser W, Fluß E, et al. EULAR revised recommendation for the management of fibromyalgia. Annals of the Rheumatic Diseases. 2017;**76**(2): 318-328

[9] Chowdhury D. Tension type headache. Annals of Indian Academy of Neurology. 2012;**15**:S83-S88

[10] Landa A, Peterson BS, Fallon BA. Somatoform pain: A developmental theory and translational research review. Psychosomatic Medicine. 2012; **74**(7):717-727

[11] Józefowicz RF, Smith JK. Diagnosis and treatment of somatoform disorders. Neurology: Clinical Practice. 2012;**2**(2): 94-102

[12] Janakiraman R, Hamilton L, Wan A. Unravelling the efficacy of antidepressants as analgesics. Australian Family Physician. 2016;**45**(3):113

[13] Brummett CM, Clauw DJ. Flipping the paradigmfrom surgery-specific to patient-driven perioperative analgesic algorithms. Anesthesiology: The Journal of the American Society of Anesthesiologists. 2015;**122**(4):731-733

[14] Ciraulo DA, Shader RI, Greenblatt DJ. Clinical pharmacology and therapeutics of antidepressants. In: Pharmacotherapy of Depression. New York City: Humana Press; 2011. pp. 33-124

[15] Hieronymus F, Lisinski A, Nilsson S, Eriksson E. Efficacy of selective serotonin reuptake inhibitors in the absence of side effects: A mega-analysis of citalopram and paroxetine in adult depression. Molecular Psychiatry. 2017. DOI: 10.1038/mp.2017.147

[16] Walker FR. A critical review of the mechanism of action for the selective serotonin reuptake inhibitors: Do these drugs possess anti-inflammatory properties and how relevant is this in the treatment of depression? Neuropharmacology. 2013;**67**:304-317

[17] Miller AH, Maletic V, Raison CL. Inflammation and its discontents: The role of cytokines in the pathophysiology

of major depression. Biological Psychiatry. 2009;**65**(9):732-741

[18] Więdłocha M, Marcinowicz P, Krupa R, Janoska-Jaździk M, Janus M, Dębowska W, et al. Effect of antidepressant treatment on peripheral inflammation markers—A meta-analysis. Progress in Neuro-Psychopharmacology and Biological Psychiatry. 2018;**80**:217-226

[19] Gałecki P, Mossakowska-Wójcik J, Talarowska M. The anti-inflammatory mechanism of antidepressants—SSRIs, SNRIs. Progress in Neuro-Psychopharmacology and Biological Psychiatry. 2018;**80**:291-294

[20] Wattiez AS, Dupuis A, Privat AM, Chalus M, Chapuy E, Aissouni Y, et al. Disruption of 5-HT2A-PDZ protein interaction differently affects the analgesic efficacy of SSRI, SNRI and TCA in the treatment of traumatic neuropathic pain in rats. Neuropharmacology. 2017;**125**:308-318

[21] Lian YN, Chang JL, Lu Q, Wang Y, Zhang Y, Zhang FM. Effects of fluoxetine on changes of pain sensitivity in chronic stress model rats. Neuroscience Letters. 2017;**651**:16-20

[22] Chen M, Hoshino H, Saito S, Yang Y, Obata H. Spinal dopaminergic involvement in the antihyperalgesic effect of antidepressants in a rat model of neuropathic pain. Neuroscience Letters. 2017;**649**:116-123

[23] Dupuis A, Wattiez AS, Pinguet J, Richard D, Libert F, Chalus M, et al. Increasing spinal 5-HT 2A receptor responsiveness mediates anti-allodynic effect and potentiates fluoxetine efficacy in neuropathic rats. Evidence for GABA release. Pharmacological Research. 2017;**118**:93-103

[24] Jung AC, Staiger T, Sullivan M. The efficacy of selective serotonin reuptake inhibitors for the management of chronic pain. Journal of General Internal Medicine. 1997;**12**(6):384-389

[25] Renoir T. Selective serotonin reuptake inhibitor antidepressant treatment discontinuation syndrome: A review of the clinical evidence and the possible mechanisms involved. Frontiers in Pharmacology. 2013;**4**:1-10

[26] Walker Z, Walker RW, Robertson MM, Stansfeld S. Antidepressant treatment of chronic tension-type headache: A comparison between fluoxetine and desipramine. Headache: The Journal of Head and Face Pain. 1998;**38**(7):523-528

[27] Max MB, Lynch SA, Muir J, Shoaf SE, Smoller B, Dubner R. Effects of desipramine, amitriptyline, and fluoxetine on pain in diabetic neuropathy. The New England Journal of Medicine. 1992;**326**(19):1250-1256

[28] Luo YL, Zhang MY, Wu WY, Li CB, Lu Z, Li QW. A randomized double-blind clinical trial on analgesic efficacy of fluoxetine for persistent somatoform pain disorder. Progress in Neuro-Psychopharmacology and Biological Psychiatry. 2009;**33**(8):1522-1525. DOI: 10.1016/j.pnpbp.2009.08.013

[29] Goldenberg D, Mayskiy M, Mossey C, Ruthazer R, Schmid C. A randomized, double-blind crossover trial of fluoxetine and amitriptyline in the treatment of fibromyalgia. Arthritis and Rheumatology. 1996;**39**(11): 1852-1859

[30] Arnold LM, Hess EV, Hudson JI, Welge JA, Berno SE, Keck PE. A randomized, placebo-controlled, double-blind, flexible-dose study of fluoxetine in the treatment of women with fibromyalgia. The American Journal of Medicine. 2002;**112**(3): 191-197

[31] Manna V, Bolino F, Cicco LD. Chronic tension-type headache, mood depression and serotonin: Therapeutic

effects of fluvoxamine and mianserine. Headache: The Journal of Head and Face Pain. 1994;**34**(1):44-49

[32] Shimodozono M, Kawahira K, Kamishita T, Ogata A, Tohgo SI, Tanaka N. Brief clinical report reduction of central poststroke pain with the selective serotonin reuptake inhibitor fluvoxamine. International Journal of Neuroscience. 2002;**112**(10):1173-1181

[33] Xiao Y, Liu J, Huang XE, Ca LH, Ma YM, Wei W, et al. Clinical study on fluvoxamine combined with oxycodone prolonged-release tablets in treating patients with moderate to severe cancer pain. Asian Pacific Journal of Cancer Prevention. 2014;**15**(23):10445-10449

[34] Riesner HJ, Zeitler C, Schreiber H, Wild A. Additional treatment in chronic pain syndrome due to hip and knee arthritis with the selective serotonin reuptake inhibitor fluvoxamine (Fevarin). Zeitschrift fur Orthopadie und Unfallchirurgie. 2008;**146**(6):742-746

[35] Kane CM, Mulvey MR, Wright S, Craigs C, Wright JM, Bennett MI. Opioids combined with antidepressants or antiepileptic drugs for cancer pain: Systematic review and meta-analysis. Palliative Medicine. 2018;**32**(1):276-286

[36] Keefe FJ, Shelby RA, Somers TJ, Varia I, Blazing M, Waters SJ, et al. Effects of coping skills training and sertraline in patients with non-cardiac chest pain: A randomized controlled study. Pain. 2011;**152**(4):730-741

[37] Lee RA, West RM, Wilson JD. The response to sertraline in men with chronic pelvic pain syndrome. Sexually Transmitted Infections. 2005;**81**(2): 147-149

[38] Engel CC, Walker EA, Engel AL, Bullis J, Armstrong A. A randomized, double-blind crossover trial of sertraline in women with chronic pelvic pain. Journal of Psychosomatic Research. 1998;**44**(2):203-207

[39] Langemark M, Olesen J. Sulpiride and paroxetine in the treatment of chronic tension-type headache. An explanatory double-blind trial. Headache: The Journal of Head and Face Pain. 1994;**34**(1):20-24

[40] Holroyd K, Labus J, O'Donell J, Cordingley G. Treating chronic tension-type headache not responding to amitriptyline hydrochloride with paroxetine hydrochloride: A pilot evaluation. Headache. 2003;**43**:999-1004

[41] Foster CA, Bafaloukos J. Paroxetine in the treatment of chronic daily headache. Headache: The Journal of Head and Face Pain. 1994;**34**(10): 587-589

[42] Dickens C, Jayson M, Sutton C, Creed F. The relationship between pain and depression in a trial using paroxetine in sufferers of chronic low back pain. Psychosomatics. 2000;**41**(6): 490-499

[43] Atkinson JH, Slater MA, Wahlgren DR, Williams RA, Zisook S, Pruitt SD, et al. Effects of noradrenergic and serotonergic antidepressants on chronic low back pain intensity. Pain. 1999; **83**(2):137-145

[44] Patkar AA, Masand PS, Krulewicz S, Mannelli P, Peindl K, Beebe KL, et al. A randomized, controlled, trial of controlled release paroxetine in fibromyalgia. The American Journal of Medicine. 2007;**120**(5):448-454

[45] Sindrup SH, Gram LF, Brøsen K, Eshøj O, Mogensen EF. The selective serotonin reuptake inhibitor paroxetine is effective in the treatment of diabetic neuropathy symptoms. Pain. 1990; **42**(2):135-144

[46] Nørregaard J, Volkmann H, Danneskiold-Samstøe B. A randomized controlled trial of citalopram in the treatment of fibromyalgia. Pain. 1995; **61**(3):445-449

[47] Anderberg UM, Marteinsdottir I, Knorring L. Citalopram in patients with fibromyalgia—A randomized, double-blind, placebo-controlled study. European Journal of Pain. 2000;**4**(1): 27-35

[48] Aragona M, Bancheri L, Perinelli D, Tarsitani L, Pizzimenti A, Conte A, et al. Randomized double-blind comparison of serotonergic (Citalopram) versus noradrenergic (Reboxetine) reuptake inhibitors in outpatients with somatoform, DSM-IV-TR pain disorder. European Journal of Pain. 2005;**9**(1): 33-38

[49] Bendtsen L, Jensen R, Olesen J. A non-selective (amitriptyline), but not a selective (citalopram), serotonin reuptake inhibitor is effective in the prophylactic treatment of chronic tension-type headache. Journal of Neurology, Neurosurgery, and Psychiatry. 1996;**61**(3):285-290

[50] Viazis N, Katopodi K, Karamanolis G, Denaxas K, Varytimiadis L, Galanopoulos M, et al. Proton pump inhibitor and selective serotonin reuptake inhibitor therapy for the management of noncardiac chest pain. European Journal of Gastroenterology and Hepatology. 2017;**29**(9):1054-1058

[51] Roohafza H, Pourmoghaddas Z, Saneian H, Gholamrezaei A. Citalopram for pediatric functional abdominal pain: A randomized, placebo-controlled trial. Neurogastroenterology and Motility. 2014;**26**(11):1642-1650

[52] Giannopoulos S, Kosmidou M, Sarmas I, Markoula S, Pelidou SH, Lagos G, et al. Patient compliance with SSRIs and gabapentin in painful diabetic neuropathy. The Clinical Journal of Pain. 2007;**23**(3):267-269

[53] Mazza M, Mazza O, Pazzaglia C, Padua L, Mazza S. Escitalopram 20 mg versus duloxetine 60 mg for the treatment of chronic low back pain. Expert Opinion on Pharmacotherapy. 2010;**11**(7):1049-1052

[54] Otto M, Bach FW, Jensen TS, Brøsen K, Sindrup SH. Escitalopram in painful polyneuropathy: A randomized, placebo-controlled, cross-over trial. Pain. 2008;**139**(2):275-283

[55] Muller JE, Wentzel I, Koen L, Niehaus DJ, Seedat S, Stein DJ. Escitalopram in the treatment of multisomatoform disorder: A double-blind, placebo-controlled trial. International Clinical Psychopharmacology. 2008;**23**(1):43-48

[56] Jaracz J, Gattner K, Moczko J, Hauser J. Comparison of the effects of escitalopram and nortriptyline on painful symptoms in patients with major depression. General Hospital Psychiatry. 2015;**37**(1):36-39

[57] Riediger C, Shuster T, Barlinn K, Maier S, Weitz J, Siepmann T. Adverse effects of antidepressants for chronic pain: A systematic review and meta-analysis. Frontiers in Neurology. 2017;**8**: 307

[58] Cegielska-Perun K, Bujalska-Zadrożny M, Tatarkiewicz J, Gąsińska E, Makulska-Nowak HE. Venlafaxine and neuropathic pain. Pharmacology. 2013;**91**(1–2):69-76

[59] Taylor K, Rowbotham MC. Venlafaxine hydrochloride and chronic pain. The Western Journal of Medicine. 1996;**165**:147-148

[60] Finnerup NB, Attal N, Haroutounian S, McNicol E, Baron R, Dworkin RH, et al. Pharmacotherapy for neuropathic pain in adults: A systematic review and meta-analysis. Lancet Neurology. 2015;**14**(2):162-173

[61] Rej S, Dew MA, Karp JF. Treating concurrent chronic low back pain and depression with low-dose venlafaxine: An initial identification of "easy-to-use"

clinical predictors of early response. Pain Medicine. 2014;**15**(7):1154-1162

[62] Richards JS, Bombardier CH, Wilson CS, Chiodo AE, Brooks L, Tate DG, et al. Efficacy of venlafaxine XR for the treatment of pain in patients with spinal cord injury and major depression: A randomized, controlled trial. Archives of Physical Medicine and Rehabilitation. 2015;**96**(4):680-689

[63] Fann JR, Bombardier CH, Richards JS, Wilson CS, Heinemann AW, Warren AM, et al. Venlafaxine extended-release for depression following spinal cord injury: A randomized clinical trial. JAMA Psychiatry. 2015;**72**(3):247-258

[64] Kus T, Aktas G, Alpak G, et al. Efficacy of venlafaxine for the relief of taxane and oxaliplatin-induced acute neurotoxicity: A single-center retrospective case-control study. Support Care Cancer. 2016;**24**(5): 2085-2091

[65] Zimmerman C, Atherton PJ, Pachman D, Seisler D, Wagner-Johnston N, Dakhil S, et al. MC11C4: A pilot randomized, placebo-controlled, double-blind study of venlafaxine to prevent oxaliplatin-induced neuropathy. Support Care Cancer. 2016; **24**(3):1071-1078

[66] Amr YM, Yousef AA. Evaluation of efficacy of the perioperative administration of venlafaxine or gabapentin on acute and chronic postmastectomy pain. The Clinical Journal of Pain. 2010;**26**(5):381-385

[67] Gallagher HC, Gallagher RM, Butler M, Buggy DJ, Henman MC. Venlafaxine for neuropathic pain in adults. The Cochrane Database of Systematic Reviews. 2015;**23**(8):CD011091

[68] Trouvin AP, Perrot S, Lloret-Linares C. Efficacy of venlafaxine in neuropathic pain: A narrative review of optimized treatment. Clinical Therapeutics. 2017;**39**(6):1104-1122

[69] Aiyer R, Barkin RL, Bhatia A. Treatment of neuropathic pain with venlafaxine: A systematic review. Pain Medicine. 2017;**18**(10):1999-2012

[70] Allen R, Sharma U, Barlas S. Clinical experience with desvenlafaxine in treatment of pain associated with diabetic peripheral neuropathy. Journal of Pain Research. 2014;**7**(23):339-351

[71] Wang CF, Russell G, Strichartz GR, Wang GK. The local and systemic actions of duloxetine in allodynia and hyperalgesia using a rat skin incision pain model. Anesthesia and Analgesia. 2015;**121**(2):532-544

[72] Stahl SM, Grady MM, Moret C, Briley M. SNRIs: Their pharmacology, clinical efficacy, and tolerability in comparison with other classes of antidepressants. CNS Spectrums. 2005; **10**(9):732-747

[73] Chapell AS, Desaiah D, Liu-Seifert H, Zhang S, Skljarveski V, Belenkov Y, et al. A double-blind, randomized, placebo-controlled study of the efficacy and safety of duloxetine for the treatment of chronic pain due to osteoarthritis of the knee. Pain Practice. 2011;**11**(1):33-41

[74] Williamson OD, Sagman D, Bruins RH, Boulay LJ, Schacht A. Antidepressants in the treatment for chronic low back pain: Questioning the validity of meta-analyses. Pain Practice. 2014;**14**(2):E33-E41

[75] Ho KY, Tay W, Yeo MC, Liu H, Yeo SJ, Chia SL, et al. Duloxetine reduces morphine requirements after knee replacement surgery. British Journal of Anaesthesia. 2010;**105**(3):371-376

[76] YaDeau JT, Brummett CM, Mayman DJ, Lin Y, Goytizolo EA, Padgett DE, et al. Duloxetine and subacute pain after

knee arthroplasty when added to a multimodal analgesic regimen: A randomized, placebo-controlled, triple-blinded trial. Anesthesiology. 2016; **125**(3):561-572

[77] Blikman T, Rienstra W, van Raaij TM, et al. Duloxetine in OsteoArthritis (DOA) study: Study protocol of a pragmatic open-label randomised controlled trial assessing the effect of preoperative pain treatment on postoperative outcome after total hip or knee arthroplasty. BMJ Open. 2016; **6**(3):e010343

[78] Dhillon S. Duloxetine: A review of its use in the management of major depressive disorder in older adults. Drugs and Aging. 2013;**30**(1):59-79

[79] Lunn MP, Hughes RA, Wiffen PJ. Duloxetine for treating painful neuropathy, chronic pain or fibromyalgia. The Cochrane Database of Systematic Reviews. 2014;**3**(1): CD007115

[80] Quilici S, Chancellor J, Löthgren M, Simon D, Said G, Le TK, et al. Meta-analysis of duloxetine vs. pregabalin in the treatment of diabetic peripheral neuropathic pain. BMC Neurology. 2009;**10**(9):6

[81] Wang ZY, Shi SY, Li SJ, Chen F, Chen H, Lin HZ, et al. Efficacy and safety of duloxetine on osteoarthritis knee pain: A meta-analysis of randomized controlled trials. Pain Medicine. 2015;**16**(7):1373-1385

[82] Lee YH, Song GG. Comparative efficacy and tolerability of duloxetine, pregabalin and milnacipran for the treatment of fibromyalgia: A Bayesian network meta-analysis of randomized trials. Rheumatology International. 2016;**36**(5):663-672

[83] Häuser W, Urrútia G, Tort S, Uceyler N, Walitt B. Serotonin and noradrenaline reuptake inhibitors (SNRIs) for fibromyalgia syndrome. Cochrane Database of Systematic Reviews. 2013;**31**(1):CD010292

[84] Puozzo C, Panconi E, Deprez D. Pharmacology and pharmacokinetcs of milnacipran. International Clinical Psychopharmacology. 2002;**17**(Supp. 1): S25-S35

[85] Briley M. Milnacipran, a well-tolerated specific serotonin and norepinephrine reuptake inhibiting antidepressant. CNS Drug Reviews. 1998;**4**:137-148

[86] Obata H, Saito S, Koizuka S, Nishikawa K, Goto F. The monoamine-mediated antiallodynic effects of intrathecally administered milnacipran, a serotonin noradrenaline reuptake inhibitor, in a rat model of neuropathic pain. Anesthesia and Analgesia. 2005; **100**:1406-1410

[87] Li C, Ji BU, Kim Y, Lee JE, Kim NK, Kim ST, et al. Electroacupuncture enhance the antiallodynic and antihyperalgesic effects of milnacipran in neuropathic rats. Anesthesia and Analgesia. 2016;**122**:1654-1662

[88] Katsuyama S, Aso H, Otowa A, Yagi T, Kishikawa Y, Komatsu T, et al. Antinociceptive effects of the serotonin and noradrenaline reuptake inhibitors milnacipran and duloxetine on vincristine-induced neuropathic pain model in mice. ISRN Pain. 2014;**2014**: 915464

[89] Cording M, Derry S, Phillips T, Moore RA, Wiffen PJ. Milnacipran for pain in fibromyalgia in adults. Cochrane Database of Systematic Reviews. 2015; **20**(10):CD008244

[90] Derry S, Phillips T, Moore RA, Wiffen PJ. Milnacipran for neuropathic pain in adults. Cochrane Database of Systematic Reviews. 2015;**6**(7): CD011789

[91] Auclaire AL, Martel JC, Assié MB, Bardin L, Heusler P, Cussac D, et al.

Levomilnacipran (F2695), a norepinephrine-preferring SNRIs : Profile in vitro and in models of depression and anxiety. Neuropharmacology. 2013;**70**:338-347

[92] Bruno A, Morabito P, Spina E, Muscatello MR. The role of levomilnacipran in the management of major depressive disorder: A comprehensive review. Current Neurpharmacology. 2016;**14**(2):191-199

[93] Montgomery SA, Mansuy L, Ruth A, Bose A, Li H, Li D. Efficacy and safety of levomilnacipran sustained release in moderate to severe major depressive disorder: A randomized, double-blind, placebo-controlled, proof-of-concept study. The Journal of Clinical Psychiatry. 2013;**74**(4):363-369

[94] Jeong H-J, Mitchell VA, Vaughan CW. Role of 5-HT1 receptor subtypes in the modulation of pain and synaptic transmission in rat spinal superficial dorsal horn. British Journal of Pharmacology. 2012;**165**(6):1956-1965

[95] Choi IS, Cho JH, Jang IS. 5-Hydroxytryptamine 1A receptors inhibit glutamate release in rat medullary dorsal horn neurons. NeuroReport. 2013;**24**(8):339-403

[96] Sałat K, Kołaczkowski M, Furgała A, Rojek A, Śniecikowska J, Varney MA, et al. Antinociceptive, antiallodynic and antihyperalgesic effects of the 5-HT1A receptor selective agonist, NLX-112 in mouse models of pain. Neuropharmacology. 2017;**125**:181-188

[97] Silvestrini B. Trazodone and the mental pain hypothesis of depression. Neuropsychobiology. 1986;**15**(Suppl 1): 2-9

[98] Mattia C, Paoletti F, Coluzzi F, Boanelli A. New antidepressants in the treatment of neuropathic pain. A review. Minerva Anestesiologica. 2002; **68**(3):105-114

[99] Pick CG, Paul D, Eison MS, Pasternak GW. Potentiation of opioid analgesia by the antidepressant nefazodone. European Journal of Pharmacology. 1992;**211**(3):375-381

[100] Ortega-Alvaro A, Acebes I, Saracíbar G, Echevarría E, Casis L, Micó JA. Effect of the antidepressant nefazodone on the density of cells expressing mu-opioid receptors in discrete brain areas processing sensory and affective dimensions of pain. Psychopharmacology. 2004;**176**(3–4): 305-311

[101] Saper JR, Lake AE, Tepper SJ. Nefazodone for chronic daily headache prophilaxis: An open-label study. Headache. 2001;**41**(5):465-474

[102] Schwartz TL, Siddiqui UA, Stahl SM. Vilazodone: A brief pharmacological and clinical review of the novel serotonin partial agonist and reuptake inhibitor. Therapeutic Advances in Psychopharmacology. 2011;**1**(3):81-87

[103] Wilson RC. The use of low-dose trazodone in the treatment of painful diabetic neuropathy. Journal of the American Podiatric Medical Association. 1999;**89**(9):468-471

[104] Ventafridda V, Caraceni A, Saita L, Bonezzi C, De Conno F, Guarise G, et al. Trazodone for deafferentation pain. Comparison with amitriptyline. Psychopharmacology. 1988;**95**(Suppl): S44-S49

[105] Goodkin K, Gullion CM, Agras WS. A randomized, double-blind, placebo-controlled trial of trazodone hydrochloride in chronic low back pain syndrome. The Journal of Clinical Psychopharmacology. 1990;**10**(4): 269-278

[106] Morillas-Arques P, Rodriguez-Lopez CM, Molina-Barea R, Rico-Villademoros F, Calandre EP. Trazodone for the treatment of

fibromyalgia: An open-label, 12-week study. BMC Musculoskeletal Disorders. 2010;**11**:204

[107] Calandre EP, Morillas-Arques P, Molina-Barea R, Rodriguez-Lopez CM, Rico-Villademoros F. Trazodone plus pregabalin combination in the treatment of fibromyalgia: A two-phase, 24-week, open-label uncontrolled study. BMC Musculoskeletal Disorders. 2011;**12**:95

[108] Davidoff G, Guarracini M, Roth E, Sliwa J, Yarkony G. Trazodone hydrochloride in the treatment of dysesthetic pain in traumatic myelopathy: A randomized, double-blind, placebo-controlled study. Pain. 1987;**29**(2):151-161

[109] Battistella PA, Ruffilli R, Cernetti R, Pettenazzo A, Baldin L, Bertoli S, et al. A placebo-controlled crossover trial using trazodone in pediatric migraine. Headache. 1993;**33**(1):36-39

[110] Frank RG, Kashani JH, Parker JC, Beck NC, Brownlee-Duffeck M, Elliott TR, et al. Antidepressant analgesia in rheumatoid arthritis. The Journal of Rheumatology. 1988;**15**(11):1632-1638

[111] Banzi R, Cusi C, Randazzo C, Sterzi R, Tedesco D, Moja L. Selective serotonin reuptake inhibitors (SSRIs) and serotonin-norepinephrine reuptake inhibitors (SNRIs) for the prevention of migraine in adults. Cochrane Database of Systematic Reviews. 2015;(5): CD011681

[112] Walitt B, Urrútia G, Nishishinya MB, Cantrell SE, Häuser W. Selective serotonin reuptake inhibitors for fibromyalgia syndrome. Sao Paulo Medical Journal. 2015;**133**(5):454

[113] Ruepert L, Quartero AO, de Wit NJ, van der Heijden GJ, Rubin G, Muris JW. Bulking agents, antispasmodics and antidepressants for the treatment of irritable bowel syndrome. Cochrane Database of Systematic Reviews. 2011;(8):CD003460

[114] Cooper TE, Wiffen PJ, Heathcote LC, Clinch J, Howard R, Krane E, et al. Antiepileptic drugs for chronic non-cancer pain in children and adolescents. The Cochrane Library. 2017;**8**:CD012536

Chapter 3

Serotonin and Emotional Decision-Making

Sara Puig Pérez

Abstract

Serotonin is one of the most important neurotransmitters involved in emotional regulation, which affect decision-making. In fact, specific genes regulate the transporter protein of serotonin, making people prone to show or not higher amygdala activation. Higher activation of specific components of the limbic system, such as amygdala, results in a higher susceptibility to make decision taking into account the environmental and emotional aspects and not only rational elaborations of the facts. It makes the response of decision-making more visceral or emotional in contrast to people with lower amygdala activation. Therefore, the importance of serotonin regulations impacts on daily and important decisions.

Keywords: serotonin, amygdala, decision-making, behavior

1. Introduction

Nowadays, decision-making has been considered as one of the most complex cognitive functions that use other superior processes such as learning, memory, and feedback sensibility [1]. Theoretically, when people have to make a decision, they are placed on a continuum of uncertainty being "complete ignorance" one pole and "certainty" the other one [2]. In this case, there are difficulties to identify the level of uncertainty (ambiguity), it could be impossible to determine the probability of gain or loss, regardless of the fact that consequences are established and clearly known. In contrast, in those situations where there is risk of uncertainty, although the consequences are stable and clearly known too, gain of fail result can be calculated. So, accordingly, decisions can be made considering as key construct the range of ambiguity or risk of uncertainty [2, 3].

The cognitive-experiential theory supported by Epstein's group [4] distinguishes between two qualitative systems of information processing in decision-making: the rational and the experiential system. The rational style of thinking is characterized by a free emotional perspective, which results in a more analytic, conscious, and effortful process. In contrast, the experiential system is based on emotions, being more automatic, effortless, imaginative, and unconscious (Epstein, 2010). Interestingly, the way people process the inputs from the environment and make decisions is considered in two ways: analytical and intuitive [5]. So, the thinking styles and the decision-making styles probably share an important background. But, decision-making is considered independently from emotional processes and brain neurochemical balance. In this chapter, we will summarize the physiological overlapping between emotional and decision-making processes and the effect of serotonin neurotransmitter on the modulation of these cognitive functions.

IntechOpen

2. Main brain regions involved in emotions and decision-making

The main brain structure involved in decision-making is the prefrontal cortex (PFC), which is considered nowadays as the main location of executing cognitive functions. In fact, it has been considered the PFC as the central key structure for cognitive functions as attention for relevant environmental stimuli, objective selection, cognitive control, planning, and monitoring performance [6, 7]. For goal-directed behavior [8] and attention [9–11], subcortical and thalamic regions are involved together with the PFC leading to a complex top-down cognitive control network. In this line, behaviors controlled by other subcortical or cortical areas out of PFC control network become habits or inflexible behaviors, mostly dependent on simple sensory motor associations [12].

It is important to take into account that the PFC region is controlled by subcortical regions involved in emotional processing such as amygdala or nucleus accumbens (NAc) [13, 14]. Both are considered key structures in processing the signal of the environmental stimuli, taking into account the emotional valence, in order to classify them as appetitive or aversive [15–17]. At the same time, PFC is connected with the orbitofrontal cortex (OFC) in voluntary choices and to the anterior cingulate cortex (ACC) in monitoring the outcome of our choices [18, 19]. These connections support the well-known effect of emotions on decision-making. In fact, lesion model studies performed with rats and nonhuman primates showed that NAc injury affects negatively response inhibition and cognitive flexibility [20–22]; meanwhile, amygdala inactivation or lesion affects the selection of relevant information, which is needed in situations with emotional value to coordinate cognitive, physiological, and behavioral responses [17, 23, 24].

Interestingly, although the dorsolateral prefrontal cortex (DLPFC) has been considered as the cognitive brain region per excellence [25, 26], there is growing evidence suggesting that DLPFC plays an important role in emotional regulation and motivated behavior [27–32].

3. Physiological overlapping between emotions and decision-making

When someone has to choose between doing something or not, he balances the immediate reward of doing it as well as the risk of incurring future negative consequences. This fact is known as the concept of willpower, which involves two brain systems' activation: the impulsive system and the reflective system [33] (see **Figure 1**). According to that, the decision-making process will mainly depend on neural substrates, which are affected by emotions because of the modulation that feelings make on neurotransmitter systems.

Main structures involved in the impulsive system are amygdala and striatum [33], which are the structures involved in short lived and very quick responses [34]. As we explained above, amygdala is responsible for attributing affective or emotional value to stimuli perceived evoked through hypothalamus and autonomic brainstem nuclei, which produce changes in internal milieu and visceral structures (e.g., *striatum* and periaqueductal gray) [35]. Amygdala plays a role in emotional decision-making, even in stimuli with an affective value learned by experience (e.g., money or drugs). It has been shown that autonomic response to the gain or loss of important amounts of money depends on the amygdala integrity [1]. In the same line, in addicts, it has been show an exacerbated autonomic response similar to monetary gains [36], which could be related to abnormal activity in the amygdala-ventral striatum system that probably would result in a heightened reward perception of the stimuli [37].

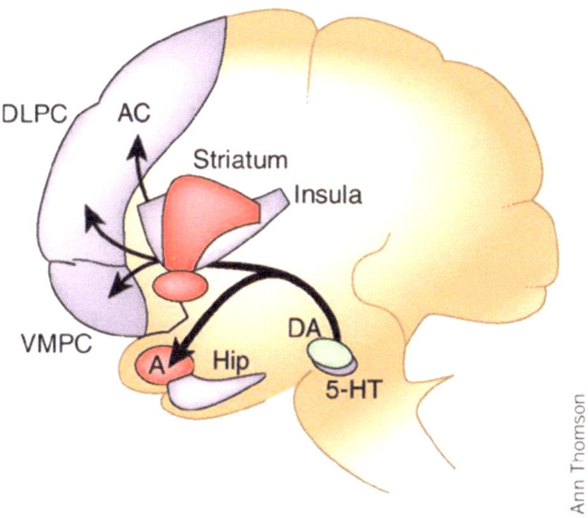

Figure 1.
Diagram from key structures involved in the impulsive (red) and reflective system (blue). A: amygdala, VMPC: ventromedial prefrontal cortex, AC: anterior cingulate, DLPC: dorsolateral prefrontal cortex, hip: hippocampus, DA: dopamine, and 5-HT: serotonin. Extracted from Ref. [33].

In contrast, the reflective system involves hippocampus, insula, AC, and PFC structures (ventromedial prefrontal cortex—VMPC, lateral orbitofrontal, inferior frontal gyrus, and DLPFC) [33]. The VMPC is crucial to induce affective states from recall to imagination, because it is responsible to reactivate the pattern of the affective state experienced in the past of reward or punishment that was developed from the brainstem nuclei [35] and to which the brain preserves the neural pattern [1]. In fact, patients with damage or people with functional abnormalities in VMPC [1] (Erns and Paulus, 2005) show impairments in decision-making. However, the normal function of the VMPC depends on the integrity of other neural systems. The insula, hippocampus, DLPFC, and somatosensory cortices need to be preserved for representing patterns of emotional or affective states and memory, but they need to preserve their integrity for the correct functioning of the VMPC too. Impairments in decision -making in patients with right parietal damage [1] and in people with functional abnormalities (Erns and Paulus, 2005), including the insula and somatosensory cortex, have been observed. Same difficulties in decision-making have been observed in patients with damage on DLPFC (Clark, Cools and Robbins, 2004). For these reasons, it can be concluded that the decision-making process depends directly on the correct functioning of other brain systems involved in memory and emotional regulation [33]. This neural overlapping suggests a functional interconnection between memory, emotions, and decision-making cognitive processes, being all connections between these brain systems are located in VMPC [1] (Clark et al., 2004).

In line with those exposed above, Rolls [38] highlights the importance of emotions in decision-making processes giving the fact that the resulting behavior of a decision is the consequence of two brain systems (see **Figure 2**). On the one hand, the emotional system is responsible for the behavior directed to reward or avoidance of punishment in terms of aptitude to natural selection [39–41]. OFC and amygdala have been related to the reward process [39], giving a reward/affective value of the stimulus processed firstly by sensorial main structures (see **Figure 2**). In contrast, the cognitive system depends on the frontoparietal network and responds to declarative knowledge and explicit goals [38]. The main contribution of Rolls [38, 39] was the stablishment of the role of lateral PFC, which is considered the hub of modulation of emotion circuity able to bias our behavior and conduct it agreeing to our explicit goals.

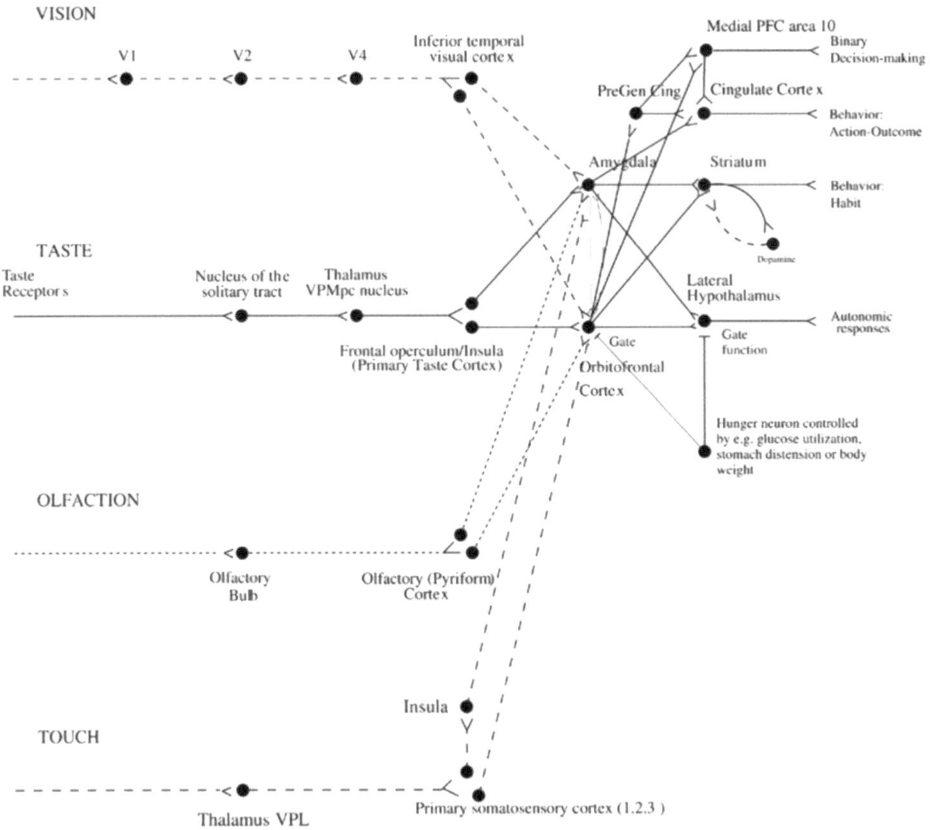

Figure 2.
Schematic diagram showing some of the connections of the taste, olfactory, somatosensory, and visual pathways in the brain. V1, primary visual (striate) cortex; V2 and V4, further cortical visual areas. PFC, prefrontal cortex. VPL, ventroposterolateral nucleus of the thalamus, which conveys somatosensory information to the primary somatosensory cortex (areas 1–3). VPMpc, ventro-postero-medial nucleus pars parvocellularis of the thalamus, which conveys the taste information to the primary taste cortex. Pregen Cing, pregenual cingulate cortex. For the purpose of description, the stages can be described as Tier 1, representing what object is present independently of reward value; Tier 2 in which reward value is represented; and Tier 3 in which decisions between stimuli of different values are taken, and in which the value is interfaced to behavioral output systems. Extracted from Rolls [39].

4. Serotonin as key neurotransmitter in emotional decision-making

Why is serotonin the hub of decision-making? Considering the neural correlation explained above which regulates emotions and decision-making, the following question to consider is which neurotransmitter regulates these networks.

Neuroanatomical studies described the brain regions mainly regulated by serotonin. It has been stated that serotoninergic neurons are located mainly in the brainstem raphe nuclei [42] and project to numerous cortical and subcortical regions. The innervation of serotoninergic neuron from this area projects to the PFC, amygdala, and NA as well as to the ventral tegmental area (VTA), which have been related to reward processes, emotions, and decision-making [13, 14, 43]. Interestingly, the medial areas from PFC play a top-down regulation from these serotoninergic regions acting as a self-feedback regulation mechanism [44]. But there is a big amount of brain regions sensible to the serotonin neurotransmitter, the effect of which depends on the receptor located in the region. Specifically, 5-HT_{1A} receptors are found mainly at hippocampus, hypothalamus, and septum which are part of the brain networks suggested as part of the different ways of decision-making process [45]. In contrast, 5-HT_{1B} receptors are found mainly on deep subcortical structures as globus pallidus, substantia nigra, ventral pallidum, and dorsal subiculum [46, 47]. The 5-HT_{2A} receptors from medial PFC are located practically exclusively on dendrites of local neuron circuits and spines of pyramidal neurons [48]. Moreover, 5-HT_{2A} receptors can be found on amygdala, hippocampus, and frontal, piriform and entorhinal cortices, as it happens with 5-HT_{2C} [49]. Finally, 5-HT_6 receptors are mainly expressed on the cerebral cortex, NA, striatum, and hippocampus [50].

Apart from neuroanatomical evidence highlighting the sensitivity from brain structures related to emotional processing and decision-making processes, one of the most important evidence of the key role of serotonin in these processes is using drugs to regulate serotonin release [51]. For example, the use of serotonin reuptake inhibitors (SSRIs) is one of the most important antidepressants used in depression and obsessive compulsive disorder [52, 53]. Interestingly, there are evidence that SSRI treatment, being effective reducing depressive symptom, improves decision-making and the sensitivity to positive feedback [54]. In this line, serotonin helps to predict future punishment [55] and affects the process of positive stimuli [54]. Increasing serotonin levels can block the uptake of serotonin released in synaptic space using, for example, SSRIs. These drugs, applied acutely, increase serotonin concentration in terminal regions, as well as reduce serotonin concentrations in raphe nuclei due to the activate 5-HT_{1A} receptor [56, 57]. Although these drugs have been considered one of the most efficient ways to increase serotonin levels, it has been observed negative effects on brain development when they have been used in young individuals [58], and to several alterations of the balance of other neurotransmitter systems [59].

Interestingly, there is a big amount of studies interested on genetic basis of serotonin, mood disorders, and cognitive functioning. Along the different genetic studies done about serotonin, it can be highlighted the role of 5-HTTLPR s-allele, which has been strongly related to anxiety traits, poor SSRI treatment response and increase of prefrontal activity [60]. C(−1019)G 5-HT1A polymorphism increases the risk of depressive disorder as well as reduces the efficacy of SSRI drug treatment [61]. And in the case of G-697C, polymorphism from 5-HT_{2C} is related to suicide [62]. Regarding 5-HT_{3A} and 5-HT_{3B}, which show a large variety of polymorphisms, they have been related to major depressive disorder [63].

Regardless of the evidence showed till now, recent studies are still trying to clarify the role of serotonin in the decision-making process. In fact, decision-making should be considered as a complex sequence of different superior process, making difficult to deeply understand the role of serotonin in each phase of brain subtasks that involves the decision-making process. In fact, the main subprocesses of decision-making with serotonin that has been considered key moderator are reversal learning, attentional set shifting, reinforce devaluation, delay discounting, and response inhibition (see [51]). In fact, Homberg [51] concluded that serotonin acts as an integrating internal and external information, affecting cognitive functioning: when serotonin is high, there is a high vigilance behavior increasing therefore the top-down PFC control which leads to an improved reversal learning, attentional set shifting, and response inhibition meanwhile decreases delay discounting. In contrast, when serotonin is low, the top-down PFC is low too leading to an impaired reversal learning, reduced attentional set shifting as well as decreased response inhibition; meanwhile, delay discounting turns high. In conclusion, serotonin regulation affects brain region functioning involved in the subtask, which depends on the decision-making process.

Author details

Sara Puig Pérez

Health Sciences Department, Valencian International University, Valencia, Spain

*Address all correspondence to: sara.puig@campusviu.es

References

[1] Bechara A. Disturbances of emotion regulation after focal brain lesions. International Review of Neurobiology. 2004;**62**(159):93

[2] Weber EU, Johnson EJ. Decisions under uncertainty: Psychological, economic, and neuroeconomic explanations of risk preference. Neuroeconomics. 2009:127-144

[3] Starcke K, Brand M. Decision making under stress: A selective review. Neuroscience and Biobehavioral Reviews. 2012;**36**(4):1228-1248

[4] Epstein S, Pacini R, Denes-Raj V, Heier H. Individual differences in intuitive-experiential and analytical-rational thinking styles. Journal of Personality and Social Psychology. 1996;**71**(2):390

[5] Hunt RG, Krzystofiak FJ, Meindl JR, Yousry AM. Cognitive style and decision making. Organizational Behavior and Human Decision Processes. 1989;**44**(3):436-453

[6] Fuster JM. The prefrontal cortex—an update: Time is of the essence. Neuron. 2001;**30**(2):319-333

[7] Miller EK. The prefrontal cortex: No simple matter. NeuroImage. 2000;**11**:447-450

[8] Floresco SB, Onge JRS, Ghods-Sharifi S, Winstanley CA. Cortico-limbic-striatal circuits subserving different forms of cost-benefit decision making. Cognitive, Affective, and Behavioral Neuroscience. 2008;**8**(4):375-389

[9] Birrell JM, Brown VJ. Medial frontal cortex mediates perceptual attentional set shifting in the rat. Journal of Neuroscience. 2000;**20**(11):4320-4324

[10] Dias R, Robbins TW, Roberts AC. Dissociation in prefrontal cortex of affective and attentional shifts. Nature. 1996;**380**(6569):69

[11] Ragozzino ME, Kesner RP. The role of rat dorsomedial prefrontal cortex in working memory for egocentric responses. Neuroscience Letters. 2001;**308**(3):145-148

[12] Miller EK. The prefontral cortex and cognitive control. Nature Reviews Neuroscience. 2000;**1**(1):59

[13] Goto Y, Grace AA. Dopaminergic modulation of limbic and cortical drive of nucleus accumbens in goal-directed behavior. Nature Neuroscience. 2005;**8**(6):805

[14] Phillips AG, Ahn S, Howland JG. Amygdalar control of the mesocorticolimbic dopamine system: Parallel pathways to motivated behavior. Neuroscience and Biobehavioral Reviews. 2003;**27**(6):543-554

[15] Faure A, Richard JM, Berridge KC. Desire and dread from the nucleus accumbens: Cortical glutamate and subcortical GABA differentially generate motivation and hedonic impact in the rat. PLoS One. 2010;**5**(6):e11223

[16] Hayes DJ, Hoang J, Greenshaw AJ. The role of nucleus accumbens shell GABA receptors on ventral tegmental area intracranial self-stimulation and a potential role for the 5-HT2C receptor. Journal of Psychopharmacology. 2011;**25**(12):1661-1675

[17] Pessoa L. Emotion and cognition and the amygdala: From "what is it?" to "what's to be done?". Neuropsychologia. 2010;**48**(12):3416-3429

[18] Rushworth MFS, Behrens TEJ, Rudebeck PH, Walton ME. Contrasting roles for cingulate and orbitofrontal cortex in decisions and social behaviour. Trends in Cognitive Sciences. 2007;**11**(4):168-176

[19] Schoenbaum G, Roesch MR, Stalnaker TA, Takahashi YK. A new perspective on the role of the orbitofrontal cortex in adaptive behaviour. Nature Reviews Neuroscience. 2009;**10**(12):885

[20] Christakou A, Robbins TW, Everitt BJ. Prefrontal cortical-ventral striatal interactions involved in affective modulation of attentional performance: Implications for corticostriatal circuit function. Journal of Neuroscience. 2004;**24**(4):773-780

[21] Clarke HF, Robbins TW, Roberts AC. Lesions of the medial striatum in monkeys produce perseverative impairments during reversal learning similar to those produced by lesions of the orbitofrontal cortex. Journal of Neuroscience. 2008;**28**(43):10972-10982

[22] Schoenbaum G, Setlow B. Lesions of nucleus accumbens disrupt learning about aversive outcomes. Journal of Neuroscience. 2003;**23**(30):9833-9841

[23] Morrison SE, Salzman CD. Re-valuing the amygdala. Current Opinion in Neurobiology. 2010;**20**(2):221-230

[24] Pessoa L, Adolphs R. Emotion processing and the amygdala: From a'low road'to'many roads' of evaluating biological significance. Nature Reviews Neuroscience. 2010;**11**(11):773

[25] D'esposito M, Postle BR. The cognitive neuroscience of working memory. Annual Review of Psychology. 2015;**66**

[26] Miller EK, Cohen JD. An integrative theory of prefrontal cortex function. Annual Review of Neuroscience. 2001;**24**(1):167-202

[27] Buhle JT, Silvers JA, Wager TD, Lopez R, Onyemekwu C, Kober H, et al. Cognitive reappraisal of emotion: A meta-analysis of human neuroimaging studies. Cerebral Cortex. 2014;**24**(11):2981-2990

[28] Fox AS, Oakes TR, Shelton SE, Converse AK, Davidson RJ, Kalin NH. Calling for help is independently modulated by brain systems underlying goal-directed behavior and threat perception. Proceedings of the National Academy of Sciences of the United States of America. 2005;**102**(11):4176-4179

[29] Frank DW, Dewitt M, Hudgens-Haney M, Schaeffer DJ, Ball BH, Schwarz NF, et al. Emotion regulation: Quantitative meta-analysis of functional activation and deactivation. Neuroscience and Biobehavioral Reviews. 2014;**45**:202-211

[30] Koenigs M, Huey ED, Calamia M, Raymont V, Tranel D, Grafman J. Distinct regions of prefrontal cortex mediate resistance and vulnerability to depression. Journal of Neuroscience. 2008;**28**(47):12341-12348

[31] Treadway MT, Buckholtz JW, Martin JW, Jan K, Asplund CL, Ginther MR, et al. Corticolimbic gating of emotion-driven punishment. Nature Neuroscience. 2014;**17**(9):1270

[32] Zaretsky M, Mendelsohn A, Mintz M, Hendler T. In the eye of the beholder: Internally driven uncertainty of danger recruits the amygdala and dorsomedial prefrontal cortex. Journal of Cognitive Neuroscience. 2010;**22**(10):2263-2275

[33] Bechara A. Decision making, impulse control and loss of willpower to resist drugs: A neurocognitive perspective. Nature Neuroscience. 2005;**8**(11):1458

[34] Büchel C, Morris J, Dolan RJ, Friston KJ. Brain systems mediating aversive conditioning: An event-related fMRI study. Neuron. 1998;**20**(5):947-957

[35] Damasio AR. Descartes' error: Emotion, rationality and the human brain. New York: Putnam (Grosset Books); 1994

[36] Bechara A. Neurobiology of decision-making: Risk and reward. Seminars in Clinical Neuropsychiatry. 2001;**6**(3):205-216

[37] Volkow ND, Fowler JS, Wang GJ. The addicted human brain viewed in the light of imaging studies: Brain circuits and treatment strategies. Neuropharmacology. 2004;**47**:3-13

[38] Rolls ET. A biased activation theory of the cognitive and attentional modulation of emotion. Frontiers in Human Neuroscience. 2013;**7**:74

[39] Rolls ET. Emotion and decision-making explained: A précis. Cortex. 2014;**59**(185):93

[40] Rolls ET. Emotion Explained. USA: Oxford University Press; 2005

[41] Rolls ET. Consciousness, decision-making and neural computation. In: Perception Action Cycle. New York, NY: Springer; 2011. pp. 287-333

[42] Azmitia EC, Segal M. An autoradiographic analysis of the differential ascending projections of the dorsal and median raphe nuclei in the rat. Journal of Comparative Neurology. 1978;**179**(3):641-667

[43] Winstanley CA, Theobald DE, Dalley JW, Cardinal RN, Robbins TW. Double dissociation between serotonergic and dopaminergic modulation of medial prefrontal and orbitofrontal cortex during a test of impulsive choice. Cerebral Cortex. 2005;**16**(1):106-114

[44] Amat J, Baratta MV, Paul E, Bland ST, Watkins LR, Maier SF. Medial prefrontal cortex determines how stressor controllability affects behavior and dorsal raphe nucleus. Nature Neuroscience. 2005;**8**(3):365

[45] Aznar S, Qian Z, Shah R, Rahbek B, Knudsen GM. The 5-HT1A serotonin receptor is located on calbindin-and parvalbumin-containing neurons in the rat brain. Brain Research. 2003;**959**(1):58-67

[46] Riad M, Garcia S, Watkins KC, Jodoin N, Doucet É, Langlois X, et al. Somatodendritic localization of 5-HT1A and preterminal axonal localization of 5-HT1B serotonin receptors in adult rat brain. Journal of Comparative Neurology. 2000;**417**(2):181-194

[47] Sari Y, Miquel MC, Brisorgueil MJ, Ruiz G, Doucet E, Hamon M, et al. Cellular and subcellular localization of 5-hydroxytryptamine1B receptors in the rat central nervous system: Immunocytochemical, autoradiographic and lesion studies. Neuroscience. 1999;**88**(3):899-915

[48] De Almeida J, Mengod G. Quantitative analysis of glutamatergic and GABAergic neurons expressing 5 HT2A receptors in human and monkey prefrontal cortex. Journal of Neurochemistry. 2007;**103**(2):475 486

[49] Miquel MC, Emerit MB, Nosjean A, Simon A, Rumajogee P, Brisorgueil MJ, et al. Differential subcellular localization of the 5-HT3-As receptor subunit in the rat central nervous system. European Journal of Neuroscience. 2002;**15**(3):449-457

[50] Gerard C, Martres MP, Lefèvre K, Miquel MC, Vergé D, Lanfumey L, et al. Immuno-localization of serotonin 5-HT6 receptor-like material in the rat central nervous system. Brain Research. 1997;**746**(1-2):207-219

[51] Homberg JR. Serotonin and decision making processes. Neuroscience and Biobehavioral Reviews. 2012;**36**(1):218-236

[52] Baumgarten HG, Grozdanovic Z. Role of serotonin in obsessive-compulsive disorder. The British Journal of Psychiatry. 1998;(Suppl. 35):13-20

[53] El Mansari M, Blier P. Mechanisms of action of current and potential pharmacotherapies of obsessive-compulsive disorder. Progress in Neuro-Psychopharmacology and Biological Psychiatry. 2006;**30**(3):362-373

[54] Bari A, Theobald DE, Caprioli D, Mar AC, Aidoo-Micah A, Dalley JW, et al. Serotonin modulates sensitivity to reward and negative feedback in a probabilistic reversal learning task in rats. Neuropsychopharmacology. 2010;**35**(6):1290

[55] Cools R, Roberts AC, Robbins TW. Serotoninergic regulation of emotional and behavioural control processes. Trends in Cognitive Sciences. 2008;**12**(1):31-40

[56] Artigas F, Bel N, Casanovas JM, Romero L. Adaptative changes of the serotonergic system after antidepressant treatments. In: Recent Advances in Tryptophan Research. Boston, MA: Springer; 1996. pp. 51-59

[57] Hajós M, Gartside SE, Sharp T. Inhibition of median and dorsal raphe neurones following administration of the selective serotonin reuptake inhibitor paroxetine. Naunyn-Schmiedeberg's Archives of Pharmacology. 1995;**351**(6):624-629

[58] Homberg JR, Schubert D, Gaspar P. New perspectives on the neurodevelopmental effects of SSRIs. Trends in Pharmacological Sciences. 2010;**31**(2):60-65

[59] Kalueff AV, Olivier JDA, Nonkes LJP, Homberg JR. Conserved role for the serotonin transporter gene in rat and mouse neurobehavioral endophenotypes. Neuroscience and Biobehavioral Reviews. 2010;**34**(3):373-386

[60] Homberg JR, Lesch KP. Looking on the bright side of serotonin transporter gene variation. Biological Psychiatry. 2011;**69**(6):513-519

[61] Albert PR, Le François B. Modifying 5-HT1A receptor gene expression as a new target for antidepressant therapy. Frontiers in Neuroscience. 2010;**4**:35

[62] Videtič A, Peternelj TT, Zupanc T, Balažic J, Komel R. Promoter and functional polymorphisms of HTR2C and suicide victims. Genes, Brain and Behavior. 2009;**8**(5):541-545

[63] Yamada K, Hattori E, Iwayama Y, Ohnishi T, Ohba H, Toyota T, et al. Distinguishable haplotype blocks in the HTR3A and HTR3B region in the Japanese reveal evidence of association of HTR3B with female major depression. Biological Psychiatry. 2006;**60**(2):192-201

[64] Floresco SB, Block AE, Maric TL. Inactivation of the medial prefrontal cortex of the rat impairs strategy set-shifting, but not reversal learning, using a novel, automated procedure. Behavioural Brain Research. 2008;**190**(1):85-96

Chapter 4

Clinical Aspects Related to Plasma Serotonin in the Horse

Katiuska Satué Ambrojo, Juan Carlos Gardon Poggi and María Marcilla Corzano

Abstract

Serotonin (5-HT) is a neurotransmitter that has important functions such as the physiological regulation of hemostasis, blood clotting, bone metabolism, cardiovascular growth, contractile activity and gastrointestinal motility, renal function, and stress and sexual behavior, among others. In this review, we consider the potential of 5-HT to contribute to the development of various pathological conditions, including metabolic, vascular, and nervous disorders in horses. The values of 5-HT in circulation are modified under common pathological conditions. Thus, laminitis, endotoxemia, surgical cramps, recurrent airway obstruction, Cushing's syndrome, central fatigue, and certain behavioral alterations such as stereotypes and other acute or chronic conditions can cause increased levels of 5-HT.

Keywords: horse, pathology, plasma serotonin

1. Introduction

Serotonin (5-hydroxytryptamine, 5-HT) is an important neurocrine messenger that is synthesized from tryptophan (TRP) by tryptophan hydroxilase in the brain and mastocytes and enterochromaffin (EC) cells in the gastrointestinal (GI) tract [1]. TRP is able to cross the blood-brain barrier and metabolize into 5-HT in the raphe nuclei within the brain stem. In the intestinal tract, 5-HT is produced by EC and to a lesser extent by serotonergic neurons and released upon mucosal stimulation. The synthesis of 5-HT is identical in the central nervous system (CNS) and in the gut, where TRP is first converted to 5-hydroxytryptophan (5-HTP) via tryptophan hydroxylase (TPH), the rate-limiting enzyme in the biosynthesis of enzyme. 5-HT is eliminated from the interstitium by 5-HT transporters on enterocytes and neurons; 5-HT overflow from the gut reaches the intestinal lumen and portal circulation. In the circulation, it is quickly removed from the plasma by uptake into platelets (PLTs) or metabolized by monoamine oxidase (MAO) into 5-hydroxyindoleacetic acid (5-HIAA) in hepatic and lung endothelial cells. Plasma 5-HT is quickly transported into the PLTs via 5-HT reuptake transporter (5-HT transporter; SERT) on the PLT membrane. PLTs accumulate, store, and release 5-HT in an analogous manner to central serotoninergic synaptosomes. The free plasma 5-HT exerts important systemic functions, modulating PLT aggregation, and has been reported to be also involved in vasomotor function [1, 2].

In GI tract, 5-HT interacts with receptors on afferent neurons, initiating peristaltic, secretion, and secretory reflexes. On the other hand, 5-HT induces smooth

IntechOpen

muscle cell contraction and proliferation but stimulates endothelial cells to release vasodilating substances and acts as a "helper agonist" of PLT aggregation [2, 3].

Measurement of 5-HT in whole blood gives a reasonable approximation of 5-HT in PLTs [4], and the free 5-HT/whole blood-5-HT (f-5HT/WB-5-HT) ratio may be a marker of PLT activation [5]. The concentration of free 5-HT is typically measured in PLT-poor plasma (PPP), produced by prolonged or high-speed centrifugation of plasma and containing <10,000 PLTs/μl [6]. Several researchers reported PPP 5-HT values in healthy horses range from 2.5 ng/ml to 90 ng/ml, with a majority varying between 3 ng/ml and 30 ng/ml [7–16]. Plasma 5-HT is the fraction which shows the equilibrium state between synthesis by EC cells, the inactivation by liver and lung by MAO and PLT uptake.

5-HT plasma concentrations in horses are subject to physiological variations such as age [17, 18], exercise [7–9], stress [19], seasonal, circadian and nycthemeral rhythms [15, 20], altitude [21], reproductive status [22], and type of anticoagulant and laboratory technique [23–25]. In addition, these factors also influence the analytical results of this neurotransmitter. Even in healthy horses, reported reference values for 5-HT are not consistent, which hampers further research into the role of 5-HT in equine diseases. One possible explanation for this inconsistency is the use of different biological samples and analytical methods for 5-HT determination. Indeed, to determine the concentrations of 5-HT in whole blood high-pressure liquid chromatography (HPLC) [7] and in serum, commercially available enzyme-linked immunosorbent assays (ELISA) or radioimmunoassays (RIAs) [19] have been used. Torfs et al. [24] showed that for accurate determination of plasma levels of 5-HT, it is essential to use PPP. It is believed that 5-HT in PPP reflects the amount of 5-HT synthesized and recently secreted in EC cells. Although ELISA [23] and HPLC [18, 21] have been used, the tandem chromatographic mass spectrometry (LC-MS/MS) method is suitable for determining the plasma reference values of 5-HT and analyzing changes in 5-HT associated with pathological conditions.

2. Role of serotonin in the equine clinic

In horses, changes in 5-HT levels are associated with gastrointestinal pathologies such as ileus, colic or endotoxemia, vascular dysfunctions such as digital hypoperfusion causing laminitis, recurrent airway obstruction and endocrine disruption such as intermediate equine pituitary dysfunction (PPID) or Cushing syndrome, and behavioral alterations such as stereotypes [26, 27].

2.1 Gastrointestinal diseases: ileus, colic, and endotoxemia

In the intestine, there are three types of cells that produce 5-HT, such as immune cells, nerve cells and EC cells [26]. Free plasma 5-HT concentration is a potential predictive parameter for postoperative ileus, since it may reflect intestinal integrity, as well as the circulatory effects associated with inflammation or endotoxemia. Therefore, 5-HT quantitation might be an aid in prognosticating the outcome in horses with postoperative colic. The knowledge of plasma 5-HT changes in colic horses is also important in the quest for an effective treatment for ileus, since certain classes of prokinetic drugs target 5-HT receptors [26]. A risk of receptor desensitization [27] might exist when these drugs are used in patients with already elevated 5-HT levels.

5-HT contractile receptors have been identified in the longitudinal and circular layers of the smooth muscle [28] and myenteric neurons of descending colon, ileum and submucosal neurons of ileum, and duodenum in horses [29, 30], in which 5-HT exerts local actions, causing the activation of intrinsic and extrinsic afferent neurons. This initiates secretory and peristaltic responses to transmit the information to the CNS [27]. While the interaction of 5-HT with 5-HT_2, 5-HT_3, and 5-HT_4 receptors stimulates contractibility, 5-HT_1 and 5-HT_7 receptors induce relaxing effects in the GI tract [31]. Among these types of receptors, 5-HT_4 exerts important control over intestinal motility. Indeed, a prokinetic effect occurs following the administration of 5-HT_4 agonists, such as Tegaserod and Mosapride [32].

In horses with intestinal hypomotility, the agonist of 5-HT_4 receptor Prucalopride can increase motor contractility of the duodenum, cecum, and colon, 30–90 minutes after oral administration. This motor activity is maximal in the duodenum, with a minimal increase in the cecum and left colon related with other intestinal segments [33]. Tegaserod is other selective agonist of 5-HT_4 receptor that induces increase in the frequency and amplitude of contractions in equine ileum and pelvic flexion [34], speeding up GI transit time and increases frequency of bowel sounds and defecation [29]. In this way, Tegaserod can offer therapeutic potential in horses suffering from impaction or paralytic ileus. Cisapride is an indirect cholinergic prokinetic agent that acts by promoting the release of acetylcholine from intramural nerve terminals (myenteric plexus) through stimulation of 5-HT_4 receptors [35]. Mosapride is other selective agonist of 5-HT_4 receptors. The use of this agonist improves gastric, jejunal, and cecal motility in horses [36]. A disadvantage of this medication is the oral administration route. This complicates the use in horses with postoperative ileus. Unfortunately, the availability of this drug is also limited. On the other hand, Tegaserod with a higher risk of cardiac events in humans has been suspected (as for cisapride), its availability is limited, as well as its application in equine practice, explaining the lack of more clinical reports. This drug is a potent dopamine D2 receptor antagonist, a moderate 5-HT_3 receptor antagonist, and 5-HT_4 receptor agonist [29] that increases the contractility of smooth muscle [37] and improved *in vivo* motility of the jejunum [36].

The final effects of 5-HT in the intestine will depend on plasma concentrations and the balance between activation and desensitization of these receptors. The concentrations of 5-HT in the intestinal mucous membrane and its association with postoperative bowel recovery may better reflect the net effects of 5-HT on intestinal motility [38].

5-HT is a very potent proinflammatory, vasoconstrictor, and immunomodulatory agent. Although Delesalle et al. [11] reported an increase in plasma concentrations of 5-HT in horses with intestinal strangulation, Ayala et al. [19] showed a decrease in serum concentration of 5-HT in horses with acute abdominal pain.

Several authors have observed higher concentrations of 5-HT after ischemia-reperfusion in the peritoneal fluid intestinal lumen and mesenteric, portal, and hepatic veins [11, 39]. Increased plasma of 5-HT can have important consequences in colic horses. It has been shown *in vitro* and *in vivo* that 5-HT is an important and very powerful vasoconstrictor agent [16]. The accumulation of 5-HT in the systemic circulation of horses that have colic may reinforce continuous intestinal ischemia. Both local lesions in the intestinal wall and the associated inflammatory and endotoxemic systemic reactions promote the development of ileus. The concentration of free 5-HT in plasma is a possible predictive parameter in cases of postoperative ileus, as it may reflect the integrity of the intestine, as well as the circulatory effects associated with inflammation or endotoxemia. For this reason, the quantification of 5-HT in horses could be an important tool to predict postsurgical evolution as a consequence of colic [26, 40].

Coagulation of circulating PLTs, as well as EC from necrotizing intestinal segments, could serve as a source of 5-HT. In horses, it is argued that intestinal ischemia makes the mucosa more permeable. This event leads to an important translocation of endotoxins and amines from the diet, among which 5-HT passes from intestinal contents to systemic circulation [28]. Indeed, Davis et al. [41] showed liver lesions in horses suffering from proximal duodenitis-jejunitis. In addition to the lesion caused by the ascending bile duct infection, these authors also propose the absorption of endotoxins or inflammatory mediators of portal circulation. In addition, hepatic hypoxia resulting from systemic inflammation and endotoxemic shock may be possible causes of liver injury.

Intestinal microorganisms are also important for the 5-HT synthesis. Yano et al. [42] estimated that 90% of peripheral 5-HT is produced by the intestinal microbiota. These authors found that in germ-free mice, the production of 5-HT was approximately 60% less in comparison to mice with normal intestinal bacteria. Indeed, when bacterial colonies were restored in the intestines of germ-free mice, 5-HT levels are recovered. Several metabolic byproducts of the intestinal microbiota are controlled by the mixture of spore-forming bacteria acting on EC to alter 5-HT production. However, bacteria are capable to produce 5-HT on their own. In fact, *Lactobacillus spp*. produce acetylcholine and GABA; *Bifidobacterium spp*. produce GABA, *Escherichia* produce norepinephrine, 5-HT, and dopamine; and *Streptococcus* and *Enterococcus* produce 5-HT. *Bacillus* species have also been shown to produce norepinephrine and dopamine [43].

Torfs et al. [24] showed that the plasma concentrations of 5-HT are significantly lower in horses with small bowel surgical colic compared to healthy animals. In addition, it was demonstrated that 5-HT concentrations remained low until at least the first morning after surgery. A previous study on horses with signs of acute colic showed significantly lower concentrations of 5-HT compared to healthy ones [19]. However, this earlier study focused on serum concentration of 5-HT used the ELISA method of analysis. This situation complicates the comparison of these results with the current ones. In contrast to these achievements, Delesalle et al. [11] indicated an increase in plasma concentrations of 5-HT, measured by HPLC, in a small group of horses undergoing small bowel surgery. In addition to the analytical differences between these studies, there are multiple possible physiological and pathological explanations for variations in the results obtained.

2.2 Vascular dysfunctions: Laminitis

The GI flora produces relatively high concentrations of dietary amines by fermenting the consumed amino acids [44, 45]. It is thought that the link between the GI system and the equine foot occurs through dietary amines.

The bacteria responsible for the fermentation of carbohydrates produce enzymes (amino acid decarboxylase) that convert free amino acids to monoamines. Then, the fermentation of large amounts of carbohydrates in the large intestine is associated with a greater number of Gram-positive bacteria (*Streptococci* and *Lactobacilli*) and elevated production of dietary amines [12]. These bacteria increase in the production of lactate, which is responsible for the decrease in intraluminal pH causing death of Gram-negative bacteria and therefore an increase in endotoxin release (lipopolysaccharide, LPS). Tryptamine is the most potent amine in the cecum. It causes vasoconstriction both *in vitro* and *in vivo* through direct activation of serotonergic receptors and 5-HT displacement of PLTs. Monoamines found in the horse's cecum could potentially induce hemodynamic disturbances in the hoof resulting in lamellar ischemia and therefore laminitis [46–50]. Besides, these monoamines present in the cecum can also be detected at much lower concentrations in blood plasma.

Leukocytes can be indirectly activated by PLTs and mainly by LPS. Therefore, endotoxin may contribute to the initiation of early inflammatory changes observed in experimental models of acute laminitis [13]. This may occur because the amines are able to mimic and potentiate the effects of the biogenic amines (5-HT, epinephrine, norepinephrine, and dopamine) in the circulation. Gut-derived amines mimic the actions of the endogenous biogenic amines by displacing 5-HT from PLTs or norepinephrine from sympathetic nerve ending or by directly activating the receptors for these amines on the vasculature [49]. The potentiation of the action of endogenous amines is produced by two processes: inhibition of absorption in endothelial cells and PLTs [47, 48] or by competition between amines derived from the intestines and endogenous amines by the metabolism of the enzyme amine oxidase [12, 47, 48].

Dietary amines are similar to substances normally produced by the body, including catecholamines and 5-HT. Since 5-HT is a potent vasoconstrictor that is mainly stored in PLTs, it helps to maintain low plasma concentrations that reduce its effects. Bailey et al. [46] reported that the absorption of 5-HT by PLTs is a saturable process in horses. The most efficient way to work is at substrate concentrations below the micromole. The noncompetitive inhibition of 5-HT absorption by other natural monoamines may result in increased plasma concentrations of 5-HT and endotoxin release. The amines present in the diet inhibit the uptake of 5-HT from the PLTs. As a result, plasma concentration of 5-HT would increase above the level at which digital vasoconstriction occurs [44, 51]. However, other peripheral blood vessels are unaffected, since digital vessels are much more sensitive to the vasoconstriction effects of 5-HT [52, 53]. Dietary amines can also cause digital vasoconstriction directly [49]. The overall result is the digital ischemia and subsequent reperfusion, which could lead to the activation of metalloproteinases.

Endotoxins act as a mechanism that triggers laminitis. These substances activate the coagulation cascade directly through the Hageman factor (factor XII, in the intrinsic coagulation pathway). They are called the contact factor because the activation occurs by contact with nonendothelial and foreign surfaces. In addition, endotoxins are the initial stage of intrinsic plasma coagulation. They also cause damage to endothelial cells and favor the addition of PLTs, thereby establishing a blood profile compatible with disseminated intravascular coagulation (DIC). As a result, peripheral vasoconstriction initially results in decreased capillary perfusion of the hoof with some degree of ischemia [47, 48].

The aggregation of PLTs and the formation of microthrombi in the capillaries of the hull contribute to maintain vascular occlusion ischemia. In addition, a potent vasoconstrictor such as thromboxane A2 is released from PLTs, which adds to increase the process. At the same time, the inflammatory response begins with the release of autacoids such as histamine, 5-HT, bradykinin, prostanoids, leukotrienes, and interleukin 1. Histamine plays a very important role in acute inflammation. It has a vasodilatory action on arterioles, but the role in inflammation is more important, since it improves the action of other mediators such as histamine and bradykinin. This results in arteriolar dilation, increased capillary permeability, and hyperalgesia. Leukotriene B4 is also involved in the passage of leukocytes to inflammatory exudate [16].

The relationship between the appearance of digital hypoperfusion and increases in plasma concentration of 5-HT is consistent. This is because PLTs-derived mediators are associated with LPS-induced laminitis. These experimental data support the use of antiPLT therapy in the prevention of laminitis related to endotoxemic diseases [16].

2.3 Endocrine diseases: pituitary pars intermedia dysfunction

Pituitary pars intermedia dysfunction (PPID) or Cushing's disease is characterized by hypertrophy and hyperplasia of the Pituitary Pars Intermedia and is argued to be due to a reduction in dopamine synthesis or degeneration of periventricular pituitary dopaminergic neurons [54]. PPID is more frequent in adult horses and result of a progressive loss of the neurotransmitter at central and peripheral level as result of the degeneration of the pineal gland [15].

The role of 5-HT in the regulation of the PPID it is not clear. Treatment of PPID in horse includes 5-HT antagonists [55], such as cyproheptadine. Antagonist of 5-HT is potent secretagogue of ACTH in pituitary rat tissue, and it was used effectively in human Cushing's disease. Cyproheptadine decreases the 5-HT-induced stimulation to the pituitary pars intermedia, the synthesis of pro-opiomelanocortin (POMC), and finally, ACTH secretion. Cyproheptadine (0.25–0.5 mg/kg PO, SID, or BID) was used for the treatment of PPID result in an improvement in clinical status and normalization of laboratory parameters within 1–2 months of treatment initiation, being effective in 28–60% of cases [56]. However, similar improvements were achieved with improved nutrition, preventive care, and management alone [57].

Additionally, Bailey et al. [46] measured peripheral plasma concentrations of 5-HT in summer, autumn, winter, and spring in clinically normal ponies and those predisposed to laminitis, and no significant differences were observed. Although light/dark differences were not investigated in the latter work, nycthemeral increases in serum 5-HT in the healthy, athletic horse have been reported [20]. Later, Haritou et al. [15] reported seasonal changes in circadian peripheral plasma concentrations of melatonin, 5-HT, dopamine, and cortisol in aged horses with PPID. Six horses and ponies with PPID were matched with six controls to test the hypothesis that aged horse responds differently to changes in season because of deficiency in melatonin production. They also examined the link between the presence or absence of the clinical signs of PPID and peripheral plasma concentration of 5-HT, dopamine, and cortisol. Results showed that the 24-h pattern of plasma melatonin concentrations during the four seasons of the year was similar in both groups, indicating that impaired melatonin output is unlikely to play a role in PPID. However, 5-HT profiles were affected by season, with lower 5-HT detected in PPID horses in the summer and winter. Although the reasons for this reduction remain unknown, enhanced conversion of 5-HT to melatonin could account, at least in part, for the lowered circulating level. The total amount of dopamine released was dependent on season and markedly lower in PPID horses versus controls. These results implicate both serotonin and dopamine in the pathogenesis of the disease [15].

2.4 Behavioral alterations: stereotypes

Most frequently observed stereotypies in domestic horses are crib biting, weaving, box walking, wind sucking, and wood chewing. However, there is no scientific consensus as to whether wood chewing is definitely a stereotypy [58]. More recently, some morphological variations of these stereotypic activities have also been identified as equine stereotypies, such as licking the environment, lip licking, sham chewing or teeth grinding, self-biting, and rubbing self, as well as locomotion stereotypies, including pawing, tail swishing, door kicking or box kicking, and head tossing/nodding [59]. The most common forms of equine stereotypies are within two general categories, oral and locomotion stereotypic behaviors.

The neurobiological consequences of equine stereotypies focus on neurotransmitter systems, specifically the serotonergic and dopaminergic pathways [59, 60]. 5-HT is implicated in the underlying pathology of stereotypies. Indeed, Lebelt et al. [61] found a trend for lower basal 5-HT levels in crib-biting compared to nonstereotypic horses, suggesting that the serotonergic system of crib-biters may differ from that of noncrib biting horses (mean 201.5 vs. 414.3 nmol/l). The precise role of 5-HT in the development or maintenance of the behavior remains unclear however, and the results obtained by these authors have yet to be confirmed or refuted through additional experimental studies of the serotonergic system in crib-biting horses.

However, blood levels of 5-HT in horses with weaving is more than triple compared to healthy horses. Thus, 5-HT is recognized as the "happiness hormone" or "pleasure hormone" [62]. It can be assumed that during the demonstration of stereotypical disorder, horses are "happy," and that repetitive disturbance is a means of increasing 5-HT levels in the blood and therefore the feeling of comfort [63].

Serotonin reuptake inhibitors (SRI) are a type of drug which acts as a reuptake inhibitor of the neurotransmitter serotonin (5-hydroxytryptamine (5-HT)) by blocking the action of the serotonin transporter (SERT). This in turn leads to increased extracellular concentrations of serotonin and, therefore, an increase in serotonergic neurotransmission. Although administration of SRIs drugs has been associated with the reduction of stereotypies, it was dubious if such medications decreased stereotypic behaviors due to general sedative effects or selectively influenced stereotypic behavior. These uncertainties were due to general sedative effects or selective influence on the type of behavior. As previously expressed, the specific role of 5-HT in the development or maintenance of behavior remains uncertain [60, 61, 64, 65], and further studies are needed to provide a more accurate interpretation of stereotypes.

Pharmacological preparations containing TRP are marketed worldwide as relaxing agents for treating excitable horses. The few studies in which TRP has been administered to horses suggest that low doses cause mild excitation. However, high doses reduce endurance capacity and cause acute hemolytic anemia when given orally due to the presence of a toxic metabolite in the intestine [65, 66]. Despite questions about its effectiveness, TRP is marketed as an equine sedative and is related to sedation, inhibition of aggression, fear, and stress.

Because TRP competes with other amino acids to bind to protein transport and cross the blood-brain barrier, researchers are now using a ratio of TRP with other large neutral amino acids (ANNALs) to estimate the production of 5-HT in the CNS [67]. Besides, previous researchers showed that the age, breed, and gender modify the response of serotoninergic system due to changes in dietary TRP [17, 68]. While all of these factors may play a role in the permeability of the blood-brain barrier, the effectiveness of supplemental TRP on 5-HT biosynthesis, it is also worth considering that these types of treatments may be most effective in horses with dysfunctioning serotoninergic system.

Although the safety of TRP doses should be confirmed, there is no evidence to suggest that a single dose is an effective analgesic in horses. In fact, given that TRP continues to be used as a tranquilizer, there is an urgent need for research to confirm its efficacy and establish a range of safe therapeutic doses. In the meantime, available data suggest that it would be unwise to rely on the TRP to calm the excitable horse. Instead, more efforts should be made to identify the underlying causes of excitability and explore other more appropriate nonpharmacological solutions. Indeed, when evaluating the use of calming supplements or drugs, it is important to consider the welfare of the horse. While calmative compounds may be beneficial in alleviating short-term stress and anxiety, the cause of such emotions should also be evaluated. Horses kept in unnatural environments, managed poorly, or asked to perform beyond their level of training may show signs of stress and anxiety. Chronic health issues, such as ulcers or lameness, may also be the culprit. Sedative drugs and supplements are often utilized to limit unwanted behaviors such as spooking, bolting, rearing, or bucking. Looking into the potential causes of unwanted behaviors should be the first step before owners turn to calming drugs or supplements. Providing more training, turnout time, or treatment for an underlying disease or condition could result in a more sustainable way to reduce a horse's unwanted behaviors and could improve welfare for the animal [65].

Equine self-mutilation syndrome (ESMS) includes glancing or biting at the flank or pectoral areas, bucking, kicking, vocalizing, rubbing, spinning, or rolling. Eight flank-biting horses with ESMS were enrolled for a behavioral study, and the effects of drugs that either stimulate or inhibit central opioid, dopamine, norepinephrine, and 5-HT neurotransmitter systems were reported. Behaviors were recorded hourly during the study and were compared with those of a saline control baseline to determine whether there were significant differences among the treatments. The suppression of ESMS activities with Buspirone (0.5 mg/kg) suggests a role for serotonergic modulation of the behavior. However, the clomipramine, a preferential 5-HT reuptake blocker, does not produce any significant effect on ESMS behavior in horses [69].

Horses with compulsive disorder may help the fluoxetine at dose of 0.25–0.5 mg/kg/day PO. Fluxetine is a selective serotonin reuptake inhibitor (*SSRI*) that increases 5-HT levels within CNS by preventing the reuptake of 5-HT at level of the presynaptic neuron. This allows 5-HT to accumulate in the synaptic cleft and affect the postsynaptic neuron. While no cases of fluoxetine-induced colic have been reported in horses being treated for behavior problems and because there are many 5-HT receptors in the gut, it is advisable to begin administering the drug at the lowest dose and increase it gradually in horses that do not improve and have not exhibited adverse effects [70].

The ability to train and control horses is an important behavioral trait for the handling and training of animals. Hori et al. [71] inform that horses carrying allele A located at c. 709G > A had a lower capacity to be handled. These results provide the first evidence that a polymorphism in a 5-HT-related gene may affect the management of horses with a partially different sex-related effect.

2.5 Recurrent airway obstruction

Based on the results reported in humans, in which PLTs contribute to the pathogenesis of allergic airway disease, Hammon et al. [14] compared PLT aggregation induced by the activating factor PLTs (PAF), thromboxane (Tx), plasma Tx, and 5-hydroxytryptamine (5-HT) production in ponies with recurrent airway obstruction (RAO) before and after antigen exposure. Plasma 5-HT was significantly higher in ponies with RAO but did not increase more by exposure to the antigen. There were also no differences between the aggregation of PLTs induced by PAF or the production of Tx or plasma Tx before or after exposure. These evidences suggest that there may be a difference between 5-HT uptake of PLTs in RAO and normal ponies. However, they do not provide evidence of PLTs activation after exposure to the antigen. This bronchoconstriction can be mediated by 5-HT. However, the effect or pathway of 5-HT may be deactivated during the development of small airway disease [72].

2.6 Central fatigue

Accordingly, the synthesis and metabolism of 5-HT in the CNS increase in response to exercise [73]. Increased 5-HT concentration in the brain is associated with central fatigue markers, such as decreased motivation, lethargy, fatigue, or loss of motor coordination [74].

It has been shown that nutritional status can alter cerebral neurochemistry, especially carbohydrates and 5-HT [75]. Therefore, it has been hypothesized that infusion of TRP may increase the ratio of fTrp (free TRP) and fTrp to BCAA while decreasing the resistance of endurance horses to treadmill. Therefore, central fatigue may limit resistance in horses, and by manipulating fTrp and BCAA, exercise capacity could be altered in a predictable way [76]. However, TRP infusion results are consistent with the central fatigue hypothesis where an increase in plasma fTrp concentration is related to the early onset of fatigue during prolonged exercise [77]. Piccione et al. [20] reported that if exercise is performed at the time of the rhythmic acrophase of the TRP (18:45, 18:16), it is likely that exercise performed at the time of the acrophase of the TRP rhythm (18:45, 18:16) affects the onset of physiological fatigue, thus turning on the body's exercise adaptation mechanisms in order to maintain better physical performance.

2.7 Other conditions

Virus of Borna's disease (BDV) can enter into the brain and infect neurons, often the limbic system. It can also remain active for long periods of time in the CNS without generating neuronal lysis. The BDV virus is unique in the order of mononegavirals because it replicates in the cell nucleus. Alterations include falling, tremor, abnormal posture, hyperactivity or hypoactivity, increased aggression, and paralysis. In some rodent species, BDV can cause mild or asymptomatic symptoms, while in other animal species such as horses, it can cause severe CNS symptoms that often lead to death. In humans, a common treatment for psychiatric illnesses such as depression or anxiety disorders is the use of SSRIs. The function of these drugs is to increase extracellular 5-HT. Interestingly, there are viruses that can exert the opposite action and reduce levels of 5-HT and therefore theoretically have an opposite effect to SSRI [78].

Equine hepatic encephalopathy is caused by direct damage to the liver or by toxins derived from the intestine that overwhelm or evade this organ. These toxins act on the CNS, giving rise to signs of encephalopathy. Secondary hepatic encephalopathy in horses occurs more often than liver failure. This may be due to megalocytic liver disease caused by ingestion of plants containing pyrrolizidine alkaloid (*Senecio, Crotalaria and Amsinckia*), Theiler's or Tyzzer's disease, cholangiohepatitis, chronic active hepatitis, liver neoplasia, toxic liver disease, and portosystemic shunts [79]. Because the liver is incapacitated, normal detoxification activities cannot be performed in all these conditions. Therefore, toxins derived from IG enter the CNS through the bloodstream.

In horses, hyperammonemia has been linked to clinical signs of encephalopathy without evidence of liver disease, which promotes the formation of Alzheimer's cells II [80]. It is also suggested that alteration of amino acid metabolism with upward regulation of aromatic amino acids and downward regulation of BCAA lead to direct neuronal inhibition secondary to effects on CNS. Alteration of *gamma aminobutyric acid* (GABA) and glutamate during liver failure plays an important role in the physiopathology of hepatic encephalopathy. Liver impairment leads to an increase in substances similar to endogenous benzodiazepines that inhibit neuronal excitation. Therefore, the most likely scenario is that there are multiple mechanisms working in synergy with each other to create signs of encephalopathy.

3. Conclusions

Serotonin is a neurotransmitter associated with important physiological, digestive, and vascular functions of the central nervous system. This review describes the involvement of serotonin in the most common pathological processes of equine clinic. Deficiencies or excesses in plasma serotonin concentrations may predispose to endocrine, vascular, or nervous disorders in the horse. Therefore, new additional studies are needed to continue showing the physiopathological mechanisms with implications of serotoninemia in other organs of the horse.

Author details

Katiuska Satué Ambrojo[1]*, Juan Carlos Gardon Poggi[2] and María Marcilla Corzano[1]

1 Department of Animal Medicine and Surgery, Cardenal Herrera University, Spain

2 Department of Animal Medicine and Surgery, Faculty of Veterinary and Experimental Sciences, Catholic University of Valencia "San Vicente Mártir", Valencia, Spain

*Address all correspondence to: ksatue@uchceu.es

References

[1] Mück-Šeler D, Pivac N. Serotonin. Periodicum Biologorum. 2011;**113**:29-41. DOI: 10.1515/tnsci-2016-0007

[2] Mohammad-Zadeh LF, Moses L, Gwaltney-Brant SM. Serotonin: A review. Journal of Veterinary Pharmacology and Therapeutics. 2008;**31**(3):187-199. DOI: 10.1111/j.1365-2885.2008.00944.x

[3] Berger M, Gray JA, Roth BL. The expanded biology of serotonin. Annual Review of Medicine. 2009;**60**:355-366. DOI: 10.1146/annurev.med.60.042307.110802

[4] Kremer HP, Goekoop JG, Van Kempen GM. Clinical use of the determination of serotonin in whole blood. Journal of Clinical Psychopharmacology. 1990;**10**(2):83-87. DOI: 10.1097/00004714-199004000-00002

[5] Hara K, Hirowatari Y, Yoshika M, Komiyama Y, Tsuka Y, Takahashi H. The ratio of plasma to whole-blood serotonin may be a novel marker of atherosclerotic cardiovascular disease. The Journal of Laboratory and Clinical Medicine. 2004;**144**:31-37. DOI: 10.1016/j.lab.2004.03.014

[6] Pappas AA, Palmer SK, Meece D, Fink LM. Rapid preparation of plasma for coagulation testing. Archives of Pathology & Laboratory Medicine. 1990;**115**:816-817 PMID: 1863193

[7] Alberghina D, Giannetto C, Piccione G. Peripheral serotoninergic response to physical exercise in athletic horses. Journal of Veterinary Science. 2010a;**11**(4):285-289. DOI: 10.4142/jvs.2010.11.4.285

[8] Alberghina D, Giannetto C, Visser EK, Ellis AD. Effect of diet on plasma tryptophan and serotonin in trained mares and geldings. The Veterinary Record. 2010b;**166**(5):133-136. DOI: 10.1136/vr.c502

[9] Alberghina D, Amorini AM, Lazzarino G. Modulation of peripheral markers of the serotoninergic system in healthy horses. Research in Veterinary Science. 2011;**90**(3):392-395. DOI: 10.1016/j.rvsc.2010.06.023

[10] Alberghina D, Giannetto C, Panzera M, Piccione G. Plasma serotonin in horses: Comparison between two different management conditions. Journal of Biological Research. 2015;**88**:5161

[11] Delesalle C, van de Walle GR, Nolten C, Ver Donck L, van Hemelrijck A, Drinkenburg W, de Bosschere H, Claes P, Deprez P, Lefere L, Torfs S, Lefebvre RA. Determination of the source of increased serotonin (5-HT) concentrations in blood and peritoneal fluid of colic horses with compromised bowel. Equine Veterinary Journal. 2008;**40**(4):326-331. DOI: 10.2746/042516408X293583

[12] Bailey SR, Katz LM, Berhane Y, Samuels T, De Brauvere N, Marr CM, Elliott J. Seasonal changes in plasma concentrations of cecum-derived amines in clinically normal ponies and ponies predisposed to laminitis. American Journal of Veterinary Research. 2003a;**64**:1132-1138. DOI: 10.2460/ajvr.2003.64.1132

[13] Bailey SR, Adair HS, Reinemeyer CR, Morgan SJ, Brooks AC, Longhofer SL, Elliott J. Plasma concentrations of endotoxin and platelet activation in the developmental stage of oligofructose-induced laminitis. Veterinary Immunology and Immunopathology. 2009;**129**(3-4):167-173. DOI: 10.1016/j.vetimm.2008.11.009

[14] Hammond A, Bailey SR, Marr CM, Cunningham FM. Platelets in equine

recurrent airway obstruction. Research in Veterinary Science. 2007;**82**(3):332-334. DOI: 10.1016/j.rvsc.2006.09.007

[15] Haritou SJ, Zylstra R, Ralli C, Turner S, Tortonese DJ. Seasonal changes in circadian peripheral plasma concentrations of melatonin, serotonin, dopamine and cortisol in aged horses with Cushing's disease under natural photoperiod. Journal of Neuroendocrino-logy. 2008;**20**(8):988-996. DOI: 10.1111/j.1365-2826.2008.01751.x

[16] Menzies-Gow NJ, Bailey SR, Katz LM, Marr CM, Elliott J. Endotoxin-induced digital vasoconstriction in horses: Associated changes in plasma concentrations of vasoconstrictor mediators. Equine Veterinary Journal. 2004;**36**(3):273-278. DOI: 10.2746/0425-164044877260

[17] Ferlazzo AM, Bruschetta G, Di Pietro P, Medica P. The influence of age on plasma serotonin levels in thoroughbred horses. Livestock Science. 2012;**147**:203-207. DOI: 10.1016/j.livsci.2012.05.010

[18] Bruschetta G, Fazio E, Cravana C, Ferlazzo AM. Effects of partial versus complete separation after weaning on plasma serotonin, tryptophan and pituitary-adrenal pattern of Anglo-Arabian foals. Livestock Science. 2017;**198**:157-161. DOI: 10.1016/j.livsci.2017.02.025

[19] Ayala I, Martos NF, Silvan G, Gutierrez-Panizo C, Clavel JG, Illera JC. Cortisol, adrenocorticotropic hormone, serotonin, adrenaline and noradrenaline serum concentrations in relation to disease and stress in the horse. Research in Veterinary Science. 2012;**93**(1):103-107. DOI: 10.1016/j.rvsc.2011.05.013

[20] Piccione G, Assenza A, Fazio F, Percipalle M, Caola G. Central fatigue and nycthemeral change of serum tryptophan and serotonin in the athletic horse. Journal of Circadian Rhythms. 2005;**3**:6. DOI: 10.1186/1740-3391-3-6

[21] Bruschetta G, Di Pietro P, Miano M, Cravana C, Ferlazzo AM. Effect of altitude on plasma serotonin levels in horses. In: Boiti C, Ferlazzo A, Gaiti A, Pugliese A, editors. Trends in Veterinary Sciences. Berlin, Heidelberg: Springer; 2013. pp. 9-13. ISBN: 978-3-642-36487-7

[22] Marcilla M, Muñoz A, Satué K. Longitudinal changes in serum catecholamines, dopamine, serotonin, ACTH and cortisol in pregnant Spanish mares. Research in Veterinary Science. 2017;**115**:29-33. DOI: 10.1016/j.rvsc.2017.01.020

[23] Torfs SC, Maes AA, Delesalle CJ, Deprez P, Croubels SM. Comparative analysis of serotonin in equine plasma with liquid chromatography—tandem mass spectrometry and enzyme-linked immunosorbent assay. Journal of Veterinary Diagnostic Investigation. 2012;**24**(6):1035-1042. DOI: 10.1177/1040638712457928

[24] Torfs SC, Maes AA, Delesalle CJ, Pardon B, Croubels SM, Deprez P. Plasma serotonin in horses undergoing surgery for small intestinal colic. The Canadian Veterinary Journal. 2015;**56**(2):178-184 PMCID: PMC4298271

[25] Satué K, Gardon JC, Marcilla M. The physiopathological features of serotonin in horses. In: Munoz M, Mckinney M, editors. Serotonin and Dopamine Receptors: Functions, Synthesis and Health Effects. Novapublisher, NY; 2018. pp. 1-43. ISBN: 978-1-53613-217-5

[26] Van Hoogmoed LM, Nieto JE, Snyder JR, Harmon FE. Survey of prokinetic use in horses with gastrointestinal injury. Veterinary Surgery. 2004;**33**:279-285. DOI: 10.1111/j.1532-950X.2004.04041.x

[27] Gershon MD, Tack J. The serotonin signaling system: From basic understanding to drug development for functional GI disorders. Gastroenterology. 2007;**132**(1):397-414. DOI: 10.1053/j.gastro.2006.11.002

[28] Delesalle C, Deprez P, Schuurkes JAJ, Lefebvre RA. Contractile effects of 5-hydroxytryptamine and 5-carboxamidotryptamine in the equine jejunum. British Journal of Clinical Pharmacology. 2006;**147**(1):23-35. DOI: 10.1038/sj.bjp.0706431

[29] Delco ML, Nieto JE, Craigmill AL, Stanley SD, Snyder JR. Pharmacokinetics and in vitro effects of tegaserod, a serotonin 5-hydroxytryptamine 4 (5-HT4) receptor agonist with prokinetic activity in horses. Veterinary Therapeutics. 2007;**8**(1):77-87 PMID: 17447227

[30] Giancola F, Rambaldi AM, Bianco F, Iusco S, Romagnoli N, Tagliavia C, Bombardi C, Clavenzani P, De Giorgio R, Chiocchetti R. Localization of the 5-hydroxytryptamine 4 receptor in equine enteric neurons and extrinsic sensory fibers. Neurogastroenterology and Motility. 2017;**29**(7). DOI: 10.1111/nmo.13045

[31] Prause AS, Stoffel MH, Portier CJ, Mevissen M. Expression and function of 5-HT7 receptors in smooth muscle preparations from equine duodenum, ileum, and pelvic flexure. Research in Veterinary Science. 2009;**87**(2):292-299. DOI: 10.1016/j.rvsc.2009.03.009

[32] Manocha M, Khan WI. Serotonin and GI disorders: An update on clinical and experimental studies. Clinical and Translational Gastroenterology. 2012;**3**:e13. DOI: 10.1038/ctg.2012.8

[33] Laus F, Fratini M, Paggi E, Faillace V, Spaterna A, Tesei B, Fettucciari K, Bassotti G. Effects of single-dose prucalopride on intestinal hypomotility in horses: Preliminary observations. Scientific Reports. 2017;7:41526. DOI: 10.1038/srep41526

[34] Lippold BS, Hildebrand J, Straub R. Tegaserod (HTF 919) stimulates gut motility in normal horses. Equine Veterinary Journal. 2004;**36**(7):622-627 PMID: 15581328

[35] Quigley EM. Cisapride: What can we learn from the rise and fall of a prokinetic? Journal of Digestive Diseases. 2011;**12**(3):147-156. DOI: 10.1111/j.1751-2980.2011.00491.x

[36] Okamura K, Sasaki N, Yamada M, Yamada H, Inokuma H. Effects of mosapride citrate, metoclopramide hydrochloride, lidocaine hydrochloride, and cisapride citrate on equine gastric emptying, small intestinal and caecal motility. Research in Veterinary Science. 2009;**86**:302-308. DOI: 10.1016/j.rvsc.2008.07.008

[37] Nieto JE, Rakestraw PC, Vatistas NJ: In vitro effects of erythromycin, lidocaine, and metoclopramide on smooth muscle from the pyloric antrum, proximal portion of the duodenum, and middle portion of the jejunum of horses. American Journal of Veterinary Research. 2000;**61**:413-419. Indexed in Pubmed PMID: 10772106

[38] Matia I, Baláz P, Jackanin S, Rybárová E, Kron I, Pomfy M, Fronek J, Ryska M. Serotonin levels in the small bowel mucosa as a marker of ischemic injury during small bowel preservation. Annals of Transplantation. 2004;**9**(3):48-51. Indexed in Pubmed PMID: 15759548

[39] Nakamura N, Hamada N, Murata R, Kobayashi A, Ishizaki N, Taira A, Sakata R. Contribution of serotonin to liver injury following canine small-intestinal ischemia and reperfusion. The Journal of Surgical Research. 2001;**99**(1):17-24. DOI: 10.1006/jsre.2001.6119

[40] Crowell MD. Role of serotonin in the pathophysiology of the irritable

bowel syndrome. British Journal of Pharmacology. 2004;**141**(8):1285-1293. DOI: 10.1038/sj.bjp.0705762

[41] Davis JL, Blikslager AT, Catto K, Jones SL. A retrospective analysis of hepatic injury in horses with proximal enteritis (1984-2002). Journal of Veterinary Internal Medicine. 2003;**17**:896-901. DOI: 10.1111/j.1939-1676.2003.tb02530.x

[42] Yano JM, Yu K, Donaldson GP, Shastri GG, Pe A, Ma L, Nagler CR, Ismagilov RF, Mazmanian SK, Hsiao EY. Indigenous bacteria from the gut microbiota regulate host serotonin biosynthesis. Cell. 2015;**161**(2):264-276. DOI: 10.1016/j.cell.2015.02.047

[43] Cryan JF, Dinan TG. Mind-altering microorganisms: The impact of the gut microbiota on brain and behaviour. Nature Reviews. Neuroscience. 2012;**13**:701-712. DOI: 10.1038/nrn3346

[44] Bailey SR, Rycroft A, Elliott J. Production of amines in equine cecal contents in an in vitro model of carbohydrate overload. Journal of Animal Science. 2002;**80**:2656-2662. DOI: 10.1093/ansci/80.10.2656

[45] Belknap JK, Black SJ. Sepsis-related laminitis. Equine Veterinary Journal. 2012;**44**(6):738-740. DOI: 10.1111/j.2042-3306.2012.00668.x

[46] Bailey SR, Cunningham FM, Elliott J. Endotoxin and dietary amines may increase plasma 5-hydroxytryptamine in the horse. Equine Veterinary Journal. 2000;**32**(6):497-504. DOI: 10.2746/042516400777584730

[47] Bailey SR, Baillon ML, Rycroft AN, Harris PA, Elliott J. Identification of equine cecal bacteria producing amines in an in vitro model of carbohydrate overload. Applied and Environmental Microbiology. 2003b;**69**:2087-2093. DOI: 10.1128/AEM.69.4.2087-2093.2003

[48] Bailey SR, Wheeler-Jones C, Elliott J. Uptake of 5-hydroxytryptamine by equine digital vein endothelial cells: Inhibition by amines found in the equine caecum. Equine Veterinary Journal. 2003c;**35**:164-169. DOI: 10.2746/042516403776114171

[49] Elliott J, Berhane Y, Bailey SR. Effects of monoamines formed in the cecum of horses on equine digital blood vessels and platelets. American Journal of Veterinary Research. 2003;**64**:1124-1131. DOI: 10.2460/ajvr.2003.64.1124

[50] Elliott J, Bailey SR. Gastrointestinal derived factors are potential triggers for the development of acute equine laminitis. The Journal of Nutrition. 2006;**136**(7):2103S-2107S. DOI: 10.1093/jn/136.7.2103S

[51] Berhane Y, Elliott J, Bailey SR. Assessment of endothelium-dependent vasodilation in equine digital resistance vessels. Journal of Veterinary Pharmacology and Therapeutics. 2006;**29**(5):387-395. DOI: 10.1111/j.1365-2885.2006.00779.x

[52] Bailey SR, Elliott J. Evidence for different 5–HT1B/1D receptors mediating vasoconstriction of equine digital arteries and veins. European Journal of Pharmacology. 1998a;**355**:175-187. DOI: 10.1016/S0014-2999(98)00520-2

[53] Bailey SR, Elliott J. Plasma 5-hydroxytryptamine constricts equine digital blood vessels in vitro: Implications for pathogenesis of acute laminitis. Equine Veterinary Journal. 1998b;**30**:124-130. DOI: 10.1111/j.2042-3306.1998.tb04471.x

[54] McFarlane D. Advantages and limitations of the equine disease, pituitary pars intermedia dysfunction as a model of spontaneous dopaminergic neurodegenerative disease. Ageing

Research Reviews. 2007;**6**(1):54-63. DOI: 10.1016/j.arr.2007.02.001

[55] Donaldson MT, McDonnell SM, Schanbacher BJ, Lamb SV, McFarlane D, Beech J. Variation in plasma adrenocorticotropic hormone concentration and dexamethasone suppression test results with season, age, and sex in healthy ponies and horses. Journal of Veterinary Internal Medicine. 2005;**19**(2):217-222. DOI: 10.1111/j.1939-1676.2005.tb02685.x

[56] Perkins G, Lamb S, Erb H, Scanbacher B, Nydam D, Divers T. Plasma adrenocorticotropin (ACTH) concentrations and clinical response in horses treated for equine Cushing's disease with cyproheptadine or pergolide. Equine Veterinary Journal. 2002;**34**(7):679-685. DOI: 10.2746/042516402776250333

[57] Schott HC II. Equine Cushing's disease: Management. In: Proceedings of the 12th International Congress of the World Equine Veterinary Association WEVA, Hyderabad, India; 2011. pp. 2-3

[58] Normando S, Meers L, Samuels WE, Faustini M, Odberg FO. Variables affecting the prevalence of behavioural problems in horses. Can riding style and other management factors be significant? Applied Animal Behaviour Science. 2011;**133**:186-198. DOI: 10.1016/j.applanim.2011.06.012

[59] McBride SD, Hemmings A. A neurologic perspective of equine stereotypy. Journal of Equine Veterinary Science. 2009;**29**:10-16. DOI: 10.1016/j.jevs.2008.11.008

[60] Wickens CL, Heleski CR. Crib-biting behavior in horses: A review. Applied Animal Behaviour Science. 2010;**128**:1-9. DOI: 10.1016/j.applanim.2010.07.002

[61] Lebelt D, Zanella AJ, Unshelm J. Physiological correlates associated with cribbing behaviour in horses: Changes in thermal threshold, heart rate, plasma beta-endorphin and serotonin. Equine Veterinary Journal. Supplement. 1998;**27**:21-27. DOI: 10.1111/j.2042-3306.1998.tb05140.x

[62] Hemmann K, Raekallio MK, Hanninen L, Pastel M, Palviainen M, Vainio O. Circadian variation in ghrelin and certain stress hormones in crib-biting horses. Veterinary Journal. 2012;**193**:97-102. DOI: 10.1016/j.tvjl.2011.09.027

[63] Binev R: Weaving horses. Etiological, clinical and paraclinical investigation. International Journal of Advanced Research 2015;3(3):629-636 http://www.journalijar.com

[64] Sarrafchi A, Blokhuis HJ. Equine stereotypic behaviors: Causation, occurrence, and prevention. Journal of Veterinary Behavior Clinical Applications and Research. 2013;**8**(5):386-394. DOI: 10.1016/j.jveb.2013.04.068

[65] Davis BP, Englea TE, Ransomb JI, Grandina T. Preliminary evaluation on the effectiveness of varying doses of supplemental tryptophan as a calmative in horses. Applied Animal Behaviour Science. 2017;**88**:34-41. DOI: 10.1016/j.applanim.2016.12.006

[66] Noble GK, Li X, Zhang D, Sillence MN. Randomised clinical trial on the effect of a single oral administration of l-tryptophan, at three dose rates, on reaction speed, plasma concentration and haemolysis in horses. Veterinary Journal. 2016;**213**:84-86. DOI: 10.1016/j.tvjl.2016.05.003

[67] Wilson AD, Badnell-Waters AJ, Bice R, Kelland A, Harris PA, Nicol CJ. The effects of diet on blood glucose, insulin, gastrin and the serum tryptophan: Large neutral amino acid ratio in foals. Veterinary Journal. 2007;**174**:139-146. DOI: 10.1016/j.tvjl.2006.05.024

[68] Alberghina D, Panzera M, Giannetto C, Piccione G. Developmental changes during the first year of life in plasma tryptophan concentration of the foal. Journal of Equine Veterinary Science. 2014;**34**:387-390. DOI: 10.1016/j.jevs.2013.07.015

[69] McDonnell SM. Practical review of self-mutilation in horses. Animal Reproduction Science. 2008;**107**(3-4):219-228. DOI: 10.1016/j.anireprosci.2008.04.012

[70] Crowell-Davis SL, Murray T. Selective serotonin reuptake inhibitors. In: Crowell-Davis SL, Murray T, editors. Veterinary Psychopharmacology. Ames, Iowa: Blackwell Publishing; 2006. pp. 80-110 ISBN: 978-0-813-80829-1

[71] Hori Y, Tozaki T, Nambo Y, Sato F, Ishimaru M, Inoue-Murayama M, Fujita K. Evidence for the effect of serotonin receptor 1A gene (HTR1A) polymorphism on tractability in thoroughbred horses. Animal Genetics. 2016;**47**(1):62-67. DOI: 10.1111/age.12384

[72] Barton AK, Wabnitz S, Kietzmann M, Ohnesorge B. Study on bronchoconstriction induced by histamine or serotonin in "Precision Cut Lung Slices" of the horse. Pferdeheilkunde. 2013;**29**(3):312-317. DOI: 10.21836/PEM20130304

[73] Chaouloff F. Physical exercise and brain monoamines: A review. Acta Physiologica Scandinavica. 1989;**137**(1):1-13. DOI: 10.1111/j.1748-1716.1989.tb08715.x

[74] Davis JM, Alderson NL, Welsh RS. Serotonin and central nervous system fatigue: Nutritional considerations. The American Journal of Clinical Nutrition. 2000;**72** (2 Suppl):573S-578S. DOI: 10.1093/ajcn/72.2.573S

[75] Fernstrom MH, Fernstrom JD. Brain tryptophan concentration and serotonin synthesis remain responsive to food consumption after the ingestion of sequential meals. The American Journal of Clinical Nutrition. 1995;**61**:312-319. DOI: 10.1093/ajcn/61.2.312

[76] Farris JW, Hincheliff KW, McKeever KH Lamb BR, Thompson DL. Effect of tryptophan and of glucose on exercise capacity of horses. Journal of Applied Physiology. 1998;**85**:807-816. DOI: 10.1152/jappl.1998.85.3.807

[77] Newsholme EA, Acworth IN, Blomstrad E. Amino acids, brain neurotransmitters and a functional link between muscle and brain that is important in sustained exercise. In: Benzi G, editor. Advances in Myochemistry. London: John Libby Eurotex; 1987. pp. 127-138 ISBN: 0861961323

[78] Richt JA, Grabner A, Herzog S. Borna disease in horses. The Veterinary Clinics of North America. Equine Practice. 2000;**16**(3):579-595. DOI: 10.1016/S0749-0739(17)30097-4

[79] Sharkey LC, DeWitt S, Stockman C. Neurologic signs and hyperammonemia in a horse with colic. Veterinary Clinical Pathology. 2006;**35**(2):254-258. DOI: 10.1111/j.1939-165X.2006.tb00126.x

[80] Mair TS, Divers TJ. Diseases of the liver and liver failure. In: Blikslager AT, White NA, Moore JM, Mair TS, editors. The Equine Acute Abdomen. NY, USA: John Wiley & Sons, Inc.; 2017. DOI: 10.1002/9781119063254.ch51